55 Advances in Polymer Science

Fortschritte der Hochpolymeren-Forschung

Solar Energy-
Phase Transfer Catalysis-
Transport Processes

With Contributions by
W. D. Comper, W. T. Ford, M. Kaneko,
B. N. Preston, M. Tomoi, A. Yamada

With 65 Figures and 21 Tables

Springer-Verlag
Berlin Heidelberg GmbH
1984

ISBN 978-3-662-15296-6   ISBN 978-3-540-38657-5 (eBook)
DOI 10.1007/978-3-540-38657-5

Library of Congress Catalog Card Number 61-642

© Springer-Verlag Berlin Heidelberg 1984
Originally published by Springer-Verlag Berlin Heidelberg New York in 1984
Softcover reprint of the hardcover 1st edition 1984

2152/3020–543210

Editors

Editorial

With the publication of Vol. 51, the editors and the publisher would like to take this opportunity to thank authors and readers for their collaboration and their efforts to meet the scientific requirements of this series. We appreciate our authors concern for the progress of Polymer Science and we also welcome the advice and critical comments of our readers.

With the publication of Vol. 51 we should also like to refer to editorial policy: *this series publishes invited, critical review articles of new developments in all areas of Polymer Science in English (authors may naturally also include works of their own)*. The responsible editor, that means the editor who has invited the article, discusses the scope of the review with the author on the basis of a tentative outline which the author is asked to provide. Author and editor are responsible for the scientific quality of the contribution; the editor's name appears at the end of it.

Manuscripts must be submitted, in content, language and form satisfactory, to Springer-Verlag. Figures and formulas should be reproducible. To meet readers' wishes, the publisher adds to each volume a "volume index" which approximately characterizes the content.

Editors and publisher make all efforts to publish the manuscripts as rapidly as possible, i.e., at the maximum, six months after the submission of an accepted paper. This means that contributions from diverse areas of Polymer Science must occasionally be united in one volume. In such cases a "volume index" cannot meet all expectations, but will nevertheless provide more information than a mere volume number.

From Vol. 51 on, each volume contains a subject index.

Editors   Publisher

Table of Contents

Solar Energy Conversion by Functional Polymers

Masao Kaneko and Akira Yamada
The Institute of Physical and Chemical Research, Wako-Shi, Saitama, 351, Japan

Advances in Polymer Science 55
© Springer-Verlag Berlin Heidelberg 1984

1 Introduction

Solar energy conversion is attracting much attention to obtain permanent and clean energy. Among the conversion systems photochemical conversion has become a rapidly growing field in recent years [1,2]. Since the solar irradiation on the earth is intermittent and unstable depending on time, season, weather, and region, the storage of the converted energy is required in order to use it in a large scale. Chemical conversion is most suited for the direct production of fuel which can be easily stored and transported.

To convert photoenergy chemically, a heterogeneous reaction system is needed to prevent energy-consuming back reactions. From that reason, the development of conversion systems composed of molecular assemblies [3-8] or polymers [9-12] has become an active center of research on solar energy conversion. The molecular assemblies such as micells, liposomes, or bilayer membranes provide microheterogeneous reaction environment, which facilitates one-directional electron flow. Macroheterogeneous conversion systems are constructed by utilizing polymers as solid phase supports or membranes. Polymer supported metal colloids and polynuclear metal complexes are the catalysts in water photolysis systems. Solar cells composed of polymer films are also important subjects.

It is to be noted that the photosynthesis in green plants [13], carried out through entire photochemical processes, is supporting all the life activities on the earth. Thus, the excellent photochemical reactions in the nature give us important informations to construct artificial photochemical conversion systems. It is encouraging to know that an anisotropic photoinduced electron flow is realized in the photosynthesis by arranging the reaction components in a bilayer membrane called Thylakoid. Molecular design of polymers and assemblies must be the key point for the construction of a synthetic conversion system.

In this review article, the functions of polymers and molecular assemblies for solar energy conversion will be described including photochemical conversion models, elemental processes for the conversion such as charge separation, electron transfer, and catalysis for water decomposition, as well as solar cells.

2 Chemical Conversion of Solar Energy

2.1 Characteristics of Solar Energy and its Chemical Conversion

Solar spectrum on the earth ranges from 250 to 2400 nm, having its maximum at 500 nm. The energy abundant ultraviolet region below 400 nm contributes only 5 %; the visible region between 400 and 800 nm occupies about half the spectrum. The conversion of this visible irradiation is therefore important.

The total energy reaching the earth amounts to 3×10^{24} J/year, which is about ten thousand times as much as the energy consumption of the mankind. For a conversion system of 10 % efficiency, 1/50 area of the desert is enough to satisfy all the energy requirements.

The conversion of the solar irradiation into electricity by solar cells seems to be the nearest goal in the present stage of research. However, considering that storage is important because of the intermittence and irregularity of irradiation, the conversion into a fuel would be desirable.

For a fuel, an electron source is needed. Water is the ultimate electron source from an economical point of view. Water photolysis is the simplest among the chemical conversion systems of solar energy. Photochemical reduction of nitrogen or carbon dioxide to produce ammonia or hydrocarbons with the electrons from water is also an attractive system of conversion.

2.2 Chemical Processes in Photosynthesis

Photosynthesis is the reduction of CO_2 by electrons from water with the help of visible irradiation producing carbohydrate and oxygen. The outline of the electrone flow is expressed by Fig. 1. The electron from water is pumped up twice by photosystems II and I (PS II and I), where chlorophyll (Chl) molecules play the main role for the excitation, energy concentration, and charge separation.

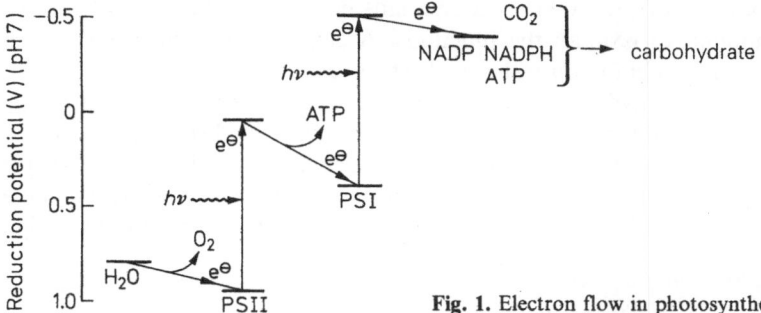

1 Chl a

The Chl (*1*), which is a magnesium complex of porphyrin containing a long alkyl chain (phytyl group), is arranged quite systematically in a Thylakoid bilayer membrane as a complex with proteins. The visible sunlight energy absorbed by the antenna Chl migrates to the reaction center Chl; the ratio of the antenna Chl amounts to 200–300 per one reaction center Chl. The excited state of the reaction

Fig. 1. Electron flow in photosynthesis

center Chl is quenched oxidatively by an acceptor molecule[1] in contact with it, followed by the reduction of the resulting Chl^+ by a donor molecule[2], thus accomplishing the charge separation. This light-harvesting and charge separation are schematically shown in Fig. 2.

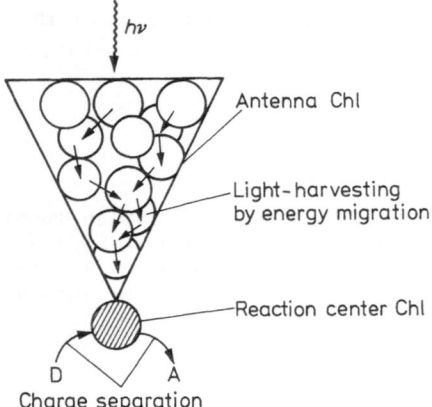

Charge separation

Fig. 2. Light harvesting and charge separation by Chl. D: Donor, A: Acceptor

For water oxidation, a redox potential of $E_0' = 2.33$ V (at pH 7, vs. NHE) is needed in the first step to abstract one electron from a water molecule (Eq. (1)). When the intermediate is stabilized on a catalyst and four electrons of two molecules of H_2O are oxidized without isolating the intermediates (so-called four-electron process), the required redox potential is only 0.82 V (Eq. (2)).

$$H_2O \rightarrow HO \cdot + e^- + H^+ \quad (E_0' = 2.33 \text{ V}) \tag{1}$$

$$2 H_2O \rightarrow O_2 + 4 e^- + 4 H^+ \quad (E_0' = 0.82 \text{ V}) \tag{2}$$

The potential level of the O_2 evolving site of the photosynthesis (see Fig. 1) ranging around 0.82 V shows that a four-electron process occurs in it. The water oxidation site of the photosynthesis contains more than four Mn ions interacting with each other, thus leading to the four-electron reaction of water to give O_2. Such a multielectron reaction leads to the generation of H_2 from proton reduction as described later in chapter 4 on water photolysis.

2.3 Photochemical Conversion Models

A model proposed for photochemical conversion of solar energy [11, 14)] is shown in Fig. 3. The system is made of a *photoreaction couple*, two kinds of *electron mediator*, and *reduction* as well as *oxidation catalysts*. It is designed to share the necessary functions among the various compounds because it would be difficult for one single compound to bear all the functions. A single component carrying out the total conversion would of course be the best system.

P$_1$ and P$_2$ are the photochemical reaction center serving also as light-harvesting unit. They can be two kinds of compounds or a single compound (P) such as a metal complex. The photoreaction center must have a strong absorption in the visible region. T_1 and T_2 are the electron mediators which take out photochemically separated charges rapidly to prevent back reactions. C_1 and C_2 are the reduction and oxidation

---

1 Pheophytin (Mg free Chl) is the likliest candidate.
2 Plastoquinone or Fe-quinone complex.

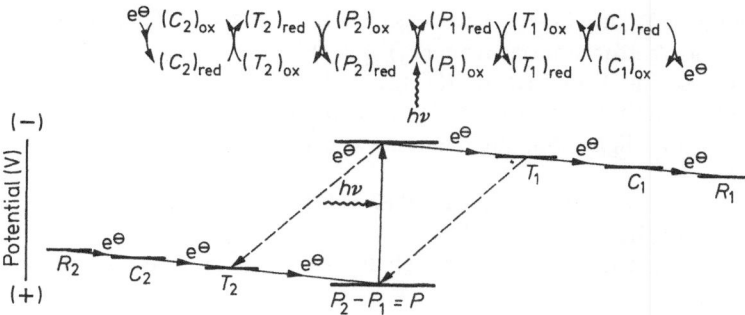

Fig. 3. A model system for photochemical conversion of solar energy. R_2: Reducing agent, R_1: Oxidizing agent; C_2: Oxidation catalyst; C_1: Reduction catalyst; T_2, T_1: Electron mediators; P_2–P_1 = P; Photoreaction center

catalytic sites, respectively. If water is to be photolyzed, it should be oxidized at C_2 to give O_2 and protons should be reduced at C_1 to give H_2 (Eqs. (3) and (4)).

$$2 H_2O \xrightarrow{C_2} O_2 + 4e^- + 4H^+ \qquad (3)$$

$$4H^+ + 4e^- \xrightarrow{C_1} 2H_2 \qquad (4)$$

In the water photolysis system, the potential of C_1 should be lower than -0.41 V and that of C_2 higher than 0.82 V. For the proton reduction, two electron process (Eq. (4), $E_0 = -0.41$ V) is much more favorable than the stepwise reaction in which the first step (Eq. (5))

$$H^+ + e^- \rightarrow H \cdot \qquad (5)$$

requires -2.52 V. Thus a multi-electron process is preferable at both the catalyst sites of water photolysis system.

In the photochemical conversion model (Fig. 3), the most serious problem is the undesired and energy-consuming *back electron transfer* (shown as dotted arrows) as well as *side electron transfer*, e.g., the electron transfer from $(C_1)_{red}$ to $(T_2)_{ox}$. It is almost impossible to prevent these undesired electron transfers, if the reactions are carried out in a homogeneous solution where all the components encounter with each other freely. In order to overcome this problem, the *use of heterogeneous conversion systems such as molecular assemblies or polymers* has attracted many researchers. The arrangement of the components on a carrier, or the separation of the T_1—C_1 sites from the T_2—C_2 ones in a heterogeneous phase must prevent the side reactions of electron transfer.

The possibilities of the two kinds of back electron transfer can be diminished to one by selecting the reaction components. When the excited state of P is quenched oxidatively by T_1, the only possible back electron transfer is from $(T_1)_{red}$, to $(P)_{ox}$. On the contrary, when P* is quenched reductively by T_2, the back electron transfer to be considered is only from $(P)_{red}$ to $(T_2)_{ox}$. In either case, the back electron transfer can be prevented by a molecular design based on reaction dynamics. For the oxidative quenching of P* by T_1, the back electron transfer from $(T_1)_{red}$ to $(P)_{ox}$ is prevented when the electron transfer from $(T_2)_{red}$ to $(P)_{ox}$ or from $(T_1)_{red}$ to $(C_1)_{ox}$ occurs much faster than the back reaction $((T_1)_{red}$ to $(P)_{ox})$.

For a model of the photoreaction center (P) to *photolyze water*, one requires:
1) Absorption of visible light with high molar extinction coefficient (ε) of over thousand.
2) A recox potential of the ground state over 0.82 V (pH 7).
3) A redox potential of the excited state below -0.41 V (pH 7).
4) A lifetime of the excited state long enough to react with acceptor or donor (redox reagent).
5) An excited state effectively quenched through electron transfer to or from redox reagent.

$Ru(bpy)_3^{2\oplus}$

Tris(2,2′-bipyridyl)ruthenium(II) complex (abbreviated to $Ru(bpy)_3^{2+}$) is the most promising candidate for the water photolysis reaction center [15]. Its absorption maximum of 452 nm in water is near the peak of solar spectrum (ca. 500 nm), and the ε value is fairly high (1.38×10^4). The redox potential ($Ru^{3+/2+}$) of $E_0' = 1.27$ V (Eq. (6)) is high enough to oxidize water. The excitation is the charge transfer from metal to ligand and the potential of the excited state (triplet) of $E_0' = -0.83$ V (Eq. (7)) is low enough to reduce protons.

$$Ru(bpy)_3^{3+} + e^- \rightarrow Ru(bpy)_3^{2+} , \quad E_0' = 1.27 \text{ V} \tag{6}$$

$$Ru(bpy)_3^{3+} + e^- \rightarrow (^3CT)Ru(bpy)_3^{2+} , \quad E_0' = -0.83 \text{ V} \tag{7}$$

These potentials theoretically allow water photolysis. However, multi-electron processes have to occur at the catalyst in order to photolyze water with this complex. The lifetime of the excited state is 650 ns, and the excited state is quenched efficiently through electron transfer with redox reagents. The conversion model with this complex is described in Chapter 4.

Other promising candidates for the P are semiconductors, which will be described in Chapter 5.

2.4 Molecular Assemblies and Polymers for Conversion Systems

Molecular assemblies such as micells or liposomes, and polymers are useful to construct solar energy conversion systems. Their effects are summarized as follows.

a) *Matrix for energy migration*:
   The energy migration in synthetic polymers is a model for light-harvesting when the trap of the captured energy is incorporated.

b) *Microenvironmental effects*:
   The electrostatic and hydrophobic microenvironment of micells and polymers facilitates charge separation from the excited state.

c) *Catalyst for multi-electron process*:
   Synthetic polymers stabilize metal colloids as important catalysts for multi-electron reactions. Polynuclear metal complexes are also efficient catalysts for multi-electron processes allowing water photolysis.

d) *Construction of heterogeneous conversion systems*:
   Molecular assemblies and polymers can separate the reaction sites and the products. Their use as a membrane or immobilized carrier leads to the photo-induced anisotropic electron flow.

e) *Utilization as membranes, films, and sheets*:
   A new type of photodiode based on electron transfer reaction of excited state is constructed by using a polymer membrane with incorporated photoactive compound as electrode coating. Stabilization of narrow bandgap semiconductors by polymer coating is a topic aiming at the utilization of liquid junction semiconductors for solar energy conversion. High conductive polymers can work as p- and n-type semiconductors for constructing solar cells made from film.

f) *Miscellaneous*:
   Polymers are used as solar concentrators, support of reaction catalysts, and electrodes for organic batteries.

   Among these subjects there have only been few works on light-harvesting. Energy migration or transfer, and excimer or exciplex formation in synthetic polymers have been studied [16, 17]. One proposed model incorporated both sensitizer as antenna and leuco crystal violet as reaction center into a polymer chain [17]. Light-harvesting will surely become important in future research, however, it is not described in the following chapters since a photoreaction center itself can act at the same time as antenna molecule.

3 Photoinduced Charge Separation and Electron Relay in Molecular and Polymer Assemblies

Photoinduced charge separation occurs when an electron is transfered from P^* to $(T_1)_{ox}$ or from $(T_2)_{red}$ to P^* (Eq. (8), (9)).

$$P + (T_1)_{ox} \xrightarrow{h\nu} P^* + (T_1)_{ox} \longrightarrow (P)_{ox} + (T_1)_{red} \qquad (8)$$

back electron transfer

$$P + (T_2)_{red} \xrightarrow{h\nu} P^* + (T_2)_{red} \longrightarrow (P)_{red} + (T_2)_{ox} \qquad (9)$$

back electron transfer

Usually the back electron transfer is so rapid that the separated charges combine again to consume the energy. When a reducing or oxidizing agent reacts rapidly with the product, however, the charge separation is completed since the back electron transfer is prevented. An example may be $Ru(bpy)_3^{2+}$ as P and methylviologen (1,1'-dimethyl-4,4'-dipyridinium dichloride, 2, abbreviated to MV^{2+}) as $(T_1)_{ox}$;

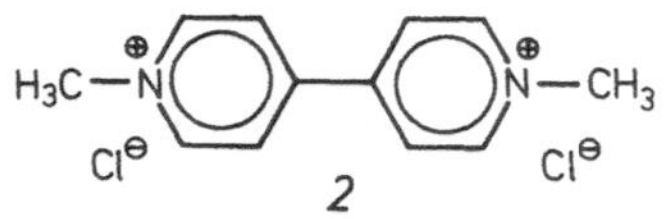

$$H_3C-\overset{\oplus}{N}\bigcirc\bigcirc\overset{\oplus}{N}-CH_3$$
$$Cl^{\ominus} \qquad\qquad\qquad Cl^{\ominus}$$
$$2$$

some sacrificial reducing agent such as EDTA or triethanolamine (THEOA) reduces rapidly the formed $Ru(bpy)_3^{3+}$ resulting in the completion of the charge separation and the accumulation of $MV^{\ddagger}$ [1,2,18] (Scheme 1).

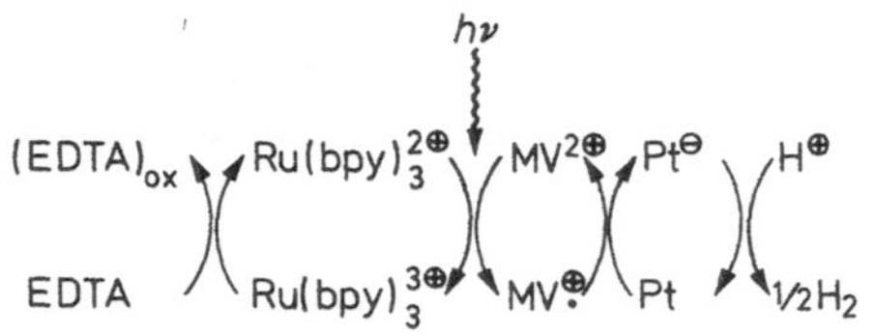

$$h\nu$$

$$(EDTA)_{ox} \diagdown \diagup Ru(bpy)_3^{2\oplus} \diagdown \diagup MV^{2\oplus} \diagdown \diagup Pt^{\ominus} \diagdown \diagup H^{\oplus}$$

$$EDTA \diagup \diagdown Ru(bpy)_3^{3\oplus} \diagup \diagdown MV^{\oplus} \diagdown Pt \diagup \diagdown {}^{1/2}H_2$$

Scheme 1

Since the reduction potential of $MV^{2+}/MV^{\ddagger}$ is low enough (-0.44 V at pH 7) to reduce protons, the presence of platinum as a catalyst in the solution containing $MV^{\ddagger}$ brings about hydrogen formation. Scheme 1 is a typical model of photo-induced charge separation and electron relay to yield H_2. It also represents the half reaction cycles of the reduction site for the photochemical conversion shown in Fig. 3.

Microenvironmental effects such as electrostatic or hydrophobic ones provided by molecular assemblies or polymers on the photoinduced charge separation and electron relay are described in this chapter together with the solid phase and macroheterogeneous photoinduced charge separation system constructed with polymer solids.

3.1 Conversion Systems on Molecular Assemblies

Molecules containing an ionic group and long alkyl chain(s) compose a molecular assembly such as a micell, a bilayer membrane, or a vesicle (liposome) (Fig. 4). [19a] These assemblies bind reaction components by hydrophobic interaction, give a high

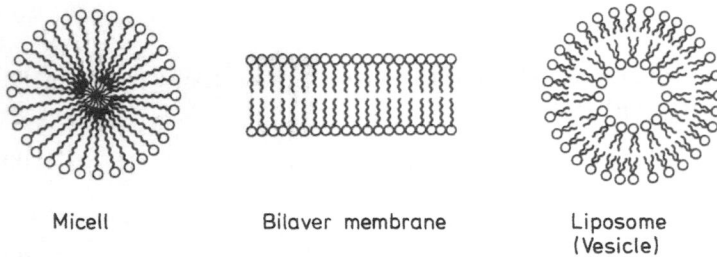

Micell   Bilayer membrane   Liposome
                                          (Vesicle)

Fig. 4. Illustrative structure of micell, bilayer membrane, and liposome (vesicle). ○ Ionic group; ～～ ; Long alkyl group

potential gradient at the interface which facilitates charge separation, and provide the heterogeneous reaction field to separate the reaction sites and products.

The ionic field of micells increases the efficiency of photoinduced charge separation. Laser flash photolysis showed a longer lifetime of the e_{aq}^- formed by irradiation of a donor molecule (D; pyrene, perylene etc.) solubilized in anionic micells such as sodium lauryl sulfate (SDS) than in a non-micell systems [19b]. This is why the e_{aq}^- is repulsed by the anionic field at the micellar surface into the bulk solution (Eq. (10)).

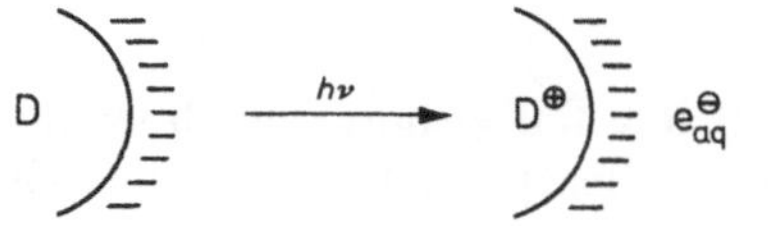

$$D \xrightarrow{h\nu} D^{\oplus} \quad e_{aq}^{\ominus} \qquad (10)$$

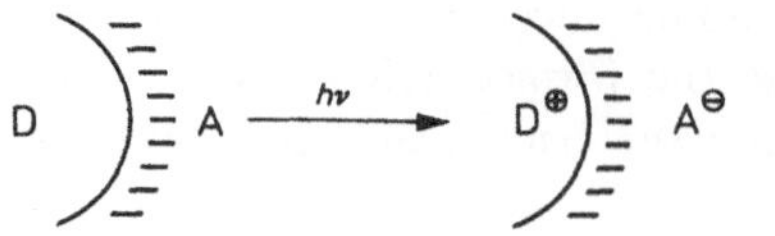

$$D \quad A \xrightarrow{h\nu} D^{\oplus} \quad A^{\ominus} \qquad (11)$$

The acceptor in the bulk solution causes a more efficient charge separation to give D^+ and A^- than non-micellar systems (Eq. (11)).

$Ru(bpy)_3^{2+}$ containing long alkyl groups at the bpy ring (3; $RuC_{12}B^{2+}$) was solubilized into micells, where the Ru complex group is located at the interface.

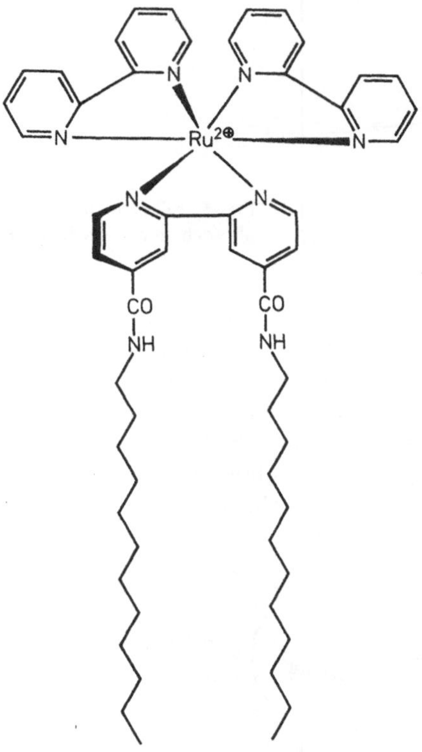

$(RuC_{12}B^{2\oplus})$

3

The irradiation of the micellar solution in the presence of dimethylaniline (DMA) which is also solubilized in the micell induced a charge separation to give $RuC_{12}B^+$ and DMA^+ (Eq. (12)).

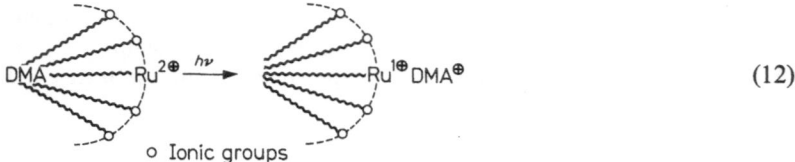

$$\tag{12}$$

o Ionic groups

The products recombine and disappear in a fairly short time, however, the lifetime of the products depends strongly on the charge of the micell [20]. The lifetime of $RuC_{12}B^+$ is the longest for the cationic micell (CTAC; cetyltrimethylammonium chloride) and the shortest for the anionic micell (SDS) (see Fig. 5). The cationic charges at the micell surface repulse the formed DMA^+ from the inside to the outside of the micell, thus preventing the back reaction between $RuC_{12}B^+$ and DMA^+.

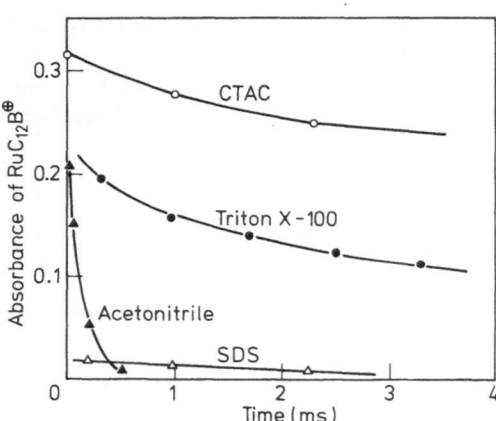

Fig. 5. Effect of micell on the lifetime of photochemically reduced $RuC_{12}B^+$

The photoinduced charge separation between N-alkyl-N'-methyl-4,4'-dipyridinium dichloride (C_nMV^{2+}; C_n = dodecyl, tetradecyl, hexadecyl, and octadecyl) and $Ru(bpy)_3^{2+}$ was also strongly affected by the presence of CTAC micells [21]. Upon reduction by $Ru(bpy)_3^{2+*}$, the viologen aquires hydrophobic properties leading to solubilization into micells (Eq. (13)). The subsequent recombination reaction is retarded by the positive surface of the micell. This decreases the rate constant of the back electron transfer at least by 500.

$$\tag{13}$$

Micell   Micell

Bilayer membranes and vesicles provide not only charged surfaces but also two phases, separating the reaction sites and products. It was first demonstrated that photoinduced electron transfer occurs from EDTA in the inner water phase of vesicles incorporated with surfactant $Ru(bpy)_3^{2+}$ to MV^{2+} in the outer water phase [22] (Eq. (14)).

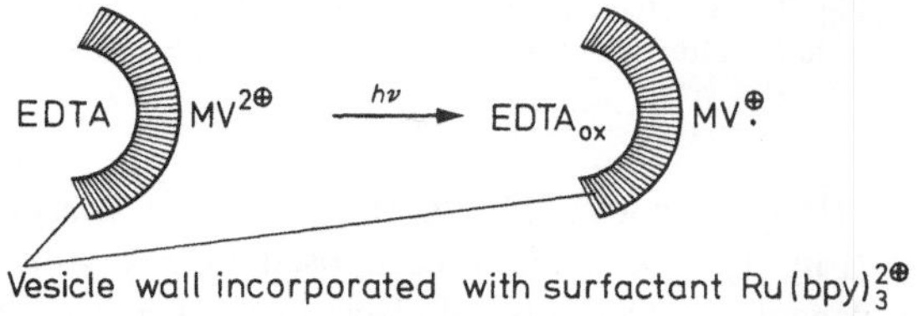

$$EDTA \quad MV^{2\oplus} \xrightarrow{h\nu} EDTA_{ox} \quad MV^{\oplus}_{\cdot} \qquad (14)$$

Vesicle wall incorporated with surfactant $Ru(bpy)_3^{2\oplus}$

In such vesicle systems, the electrons are transported through the membrane. Electron carriers such as quinones or alloxazines in the vesicle wall enhance remarkably the rate of photoinduced charge separation. The vesicle system shown in Fig. 6 contains the surfactant Zn-porphyrine complex ($ZnC_{12}TPyP$) in the wall [23].

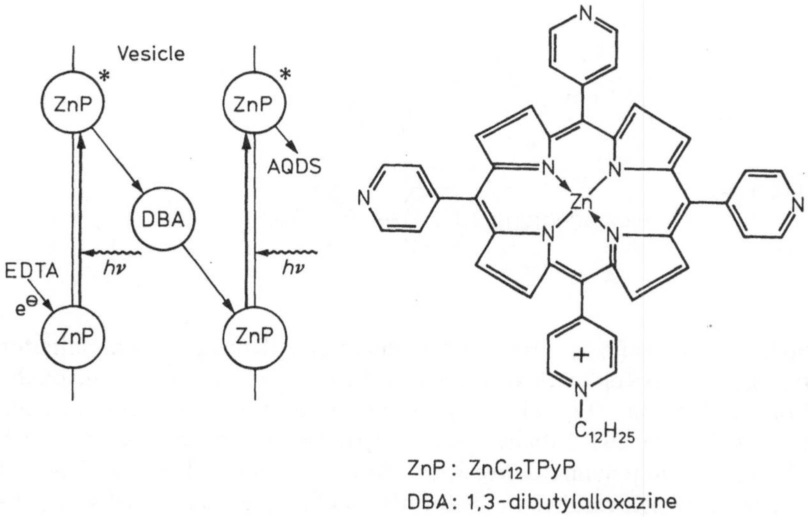

ZnP: $ZnC_{12}TPyP$
DBA: 1,3-dibutylalloxazine

Fig. 6. Vesicle system for photochemical charge separation

The incorporation of 1,3-dibutylalloxazine (DBA) as an electron carrier in the vesicle wall increased the accumulation rate of the reduced disodium 9,10-anthraqui-none-2,6-disulfate ($AQDSH_2$) by 7 times. In the system of Fig. 6, the electron is pumped up in two steps at the inside as well as outside of the vesicle wall, however, since the photoreaction centers (surfactant ZnP) are the same for both the walls, the amount of energy acquired by the two steps excitation is equal to that by one step excitation. The incorporation of different photoreaction centers inside and

M. Kaneko and A. Yamada

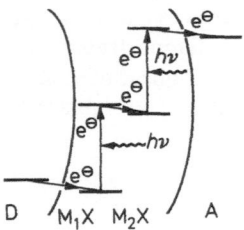

Fig. 7. Two step pumping of electron at inside and outside of vesicle wall. D: Donor; A: Acceptor; M_1X, M_2X: Sensitizers

outside the vesicle walls (Fig. 7) would lead to the two steps energy aquirement like the photosynthesis (Fig. 1). The vesicles composed of egg yolk lecithin and incorporated with methyleneblue (MB) induced electron transport photochemically from the sodium ascorbate in the inner aqueous phase to the ferri cyanide in the outer phase [24].

These molecular assemblies are unfortunately not stable enough to construct practical solar energy conversion systems. *Vesicles* composed of *polymerizable monomers* (e.g., *4*, *5*) were polymerized to give polymeric vesicles having enhanced stability [25, 26].

$$CH_2=CH(CH_2)_8COO(CH_2)_2$$
$$CH_2=CH(CH_2)_8COO(CH_2)_2 \quad NCO(CH_2)_2 \overset{\oplus}{N} \bigcirc \text{—} \bigcirc \overset{\oplus}{N}CH_3 \quad Br^{\ominus}, I^{\ominus}$$
4

$$CH_3(CH_2)_{14}COO(CH_2)_2$$
$$CH_3(CH_2)_{14}COO(CH_2)_2 \quad NCOCH=CHCOO(CH_2)_2 \overset{\oplus}{N} \bigcirc \text{—} \bigcirc \overset{\oplus}{N}CH_3 \quad Br^{\ominus}, I^{\ominus}$$
5

On sonication, surfactants (*4*, *5*) form vesicles which are polymerized by an initiator or by UV irradiation across either their bilayers or their head groups depending on the position of the double bond (Fig. 8). The polymeric vesicles are stable for extended periods even in 25% C_2H_5OH. Efficient charge separation has been realized in such chemically disymmetrical polymerized vesicles. Photoexcitation of $Ru(bpy)_3^{2+}$ placed on the outside of the vesicle resulted in the formation of long-lived reduced viologens on the inside.

Polymeric microemulsions ranging in size from 200 to 400 Å in diameter were prepared by polymerizing an oil in water microemulsion consisting of cetyltrimethyl-ammonium bromide, styrene, hexanol in water by an initiator or by irradiation. The polymerized microemulsions were studied by electron micrography. These microemulsions provide two sites for reactants, i.e., hydrophobic microenvironment in the surfactant spherical shell and ionic surface of the particle. The electron transfer from surface located dimethylaniline to sphere immobilized excited pyrene was investigated, and different kinetics were observed from those of micells [27]. The pyrene was bound at the two different sites. The excited guest molecules in the ionic surface region are readily quenched by the quencher (dimethylaniline), while

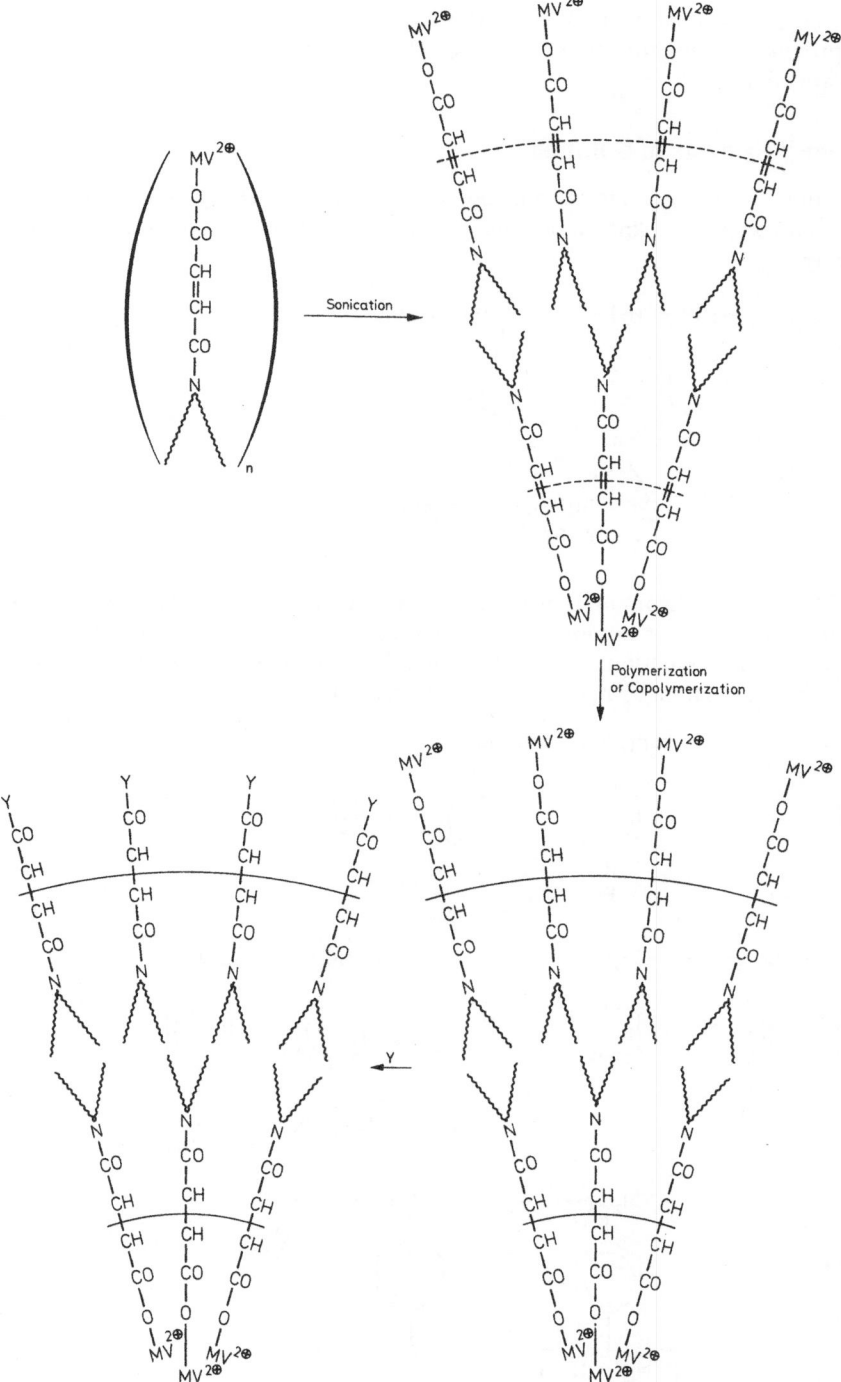

Fig. 8. Proposed formation of polymeric surfactant vesicles in which the outer redox-active viologen groups have been removed by nucleophilic cleavage of the ester bond

those in the polymer core are relatively unaffected. It was proposed that these systems provide vesicles for studying photoinduced reactions where the effects of diffusion are minimized.

3.2 Polymer Systems in Solution

In this section, $Ru(bpy)_3^{2+}$ and its polymer analogoues are mainly described, since the Ru complex is best known as the photocatalyst that can photolyze water theoretically.

Bis(2,2′-bipyridyl)ruthenium(II) was anchored onto poly(4-vinylpyridine) (PVP) (6) [28,29], but the polymer complex is not suitable as photocatalyst, because it is ⸱ susceptible to photoaquation. *A polymer complex containing $Ru(bpy)_3^{2+}$ pendant groups* was first prepared by reaction of polystyrene (PSt) as shown in Eq. (15) [30].

(15)

The polymer complex, Ru(PSt-bpy) (bpy)$_2^{2+}$, is stable under irradiation, and shows almost the same sensitizing ability in the photoreduction of MV^{2+} in solution [30, 31]. The solubility of the polymer complex, quite different from the monomeric Ru(bpy)$_3^{2+}$ allows its use as solid phase catalyst in water as described in the next section [31].

4-Methyl-4'-vinyl-2,2'-bipyridyl (7; Vbpy) was prepared and the Ru complex of its trichlorosilylethyl derivative was coated on n-SnO$_2$ by condensation of the surface hydroxyl groups [32]. Anodic photocurrent obtained at the Ru(bpy)$_3^{2+}$ bound n-SnO$_2$ semiconductor electrode was about twice than that obtained at the bare SnO$_2$ dipped in 0.1 N H$_2$SO$_4$ solution of Ru(bpy)$_3^{2+}$.

Vbpy was copolymerized with styrene to give the copolymer, P(St-Vbpy), which was then treated with cis-Ru(bpy)$_2$Cl$_2$ in xylene/n-butylalcohol = 1/4 (v/v) under reflux to form polymer pendant Ru(bpy)$_3^{2+}$ (8; Ru[P(St-Vbpy)] (bpy)$_2^{2+}$) [10].

The solubility of the complex changes by being anchored on a polymer. The monomeric Ru(bpy)$_3^{2+}$ is well soluble in water, but insoluble in organic solvents such as benzene or chloroform. The polymer complex (8) is, on the contrary, quite insoluble in water, but well soluble in benzene and chloroform. The absorption and emission

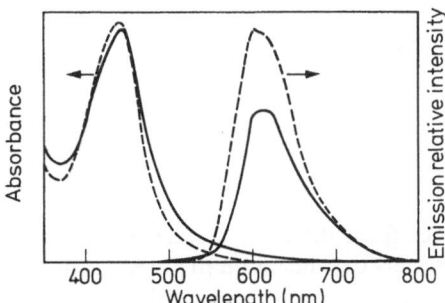

Fig. 9. Absorption and emission spectra of Ru[P(St-Vbpy)](bpy)$_2^{2+}$ (—) and Ru(bpy)$_3^{2+}$ (...) in DMF

spectra of *8* are very similar to $Ru(bpy)_3^{2+}$, although the relative emission intensity is somewhat weaker than $Ru(bpy)_3^{2+}$ (Fig. 9).

In the photoreaction system of $Ru(bpy)_3^{2+}$ and MV^{2+}, the quenching of the emission from $Ru(bpy)_3^{2+*}$ is due to the electron transfer from the excited state to MV^{2+}. The rate of the electron transfer from $Ru(bpy)_3^{2+*}$ to MV^{2+} is therefore represented by the quenching rate constant (k_q). This is calculated from the Stern-Volmer constant (k_{sv}) which is the slope of the plots of relative emission intensity vs. $[MV^{2+}]$, and the lifetime of $Ru(bpy)_3^{2+*}$ (τ) (Eqs. (16 and 17)),

$$I_0/I = 1 + k_{sv}[MV^{2+}] \tag{16}$$

$$k_q = k_{sv}/\tau \tag{17}$$

where I_0 and I are the relative emission intensities in the absence and presence of MV^{2+}, respectively. The values τ, k_{sv} and k_q for the polymer complex (*8*) and $Ru(bpy)_3^{2+}$ in homogeneous solution of $DMF/H_2O = 9/1$ are shown in Table 1 [10]. The values k_q are not so much different for both the complexes, i.e., the excited state of the polymer complex has almost the same activity as the monomeric one.

Table 1. Rate constant of electron transfer (k_q) from excited Ru complex to MV^{2+} for polymer pendant $Ru(bpy)_3^{2+}$ in $DMF/H_2O = 9/1$

Complex	τ(ns)	k_{sv} (M^{-1})	k_q (M$^{-1} \cdot$ s^{-1})
$Ru[P(Vbpy\text{-}St)](bpy)_2^{2+}$	434	245	5.65×10^8
$Ru(bpy)_3^{2+}$	350	234	6.69×10^8

Measured at the MV^{2+} concentration below 1 mM

The ionic domain around the complex affects the rate of photochemical reaction by attracting or repulsing the substrate. The polymer pendant $Ru(bpy)_3^{2+}$ was prepared from a poly(Vbpy-co-acrylic acid) (*9*; $Ru[P(Vbpy\text{-}AA)] (bpy)_2^{2+}$) [33].

9

N⌣N = 2,2'-bipyridil

The rate constant (k_q) of the electron transfer from the excited state of the water soluble polymer complex (*9*) to MV^{2+} was strongly dependent on pH of the solution. The pH dependences of k_q for both *9* and $Ru(bpy)_3^{2+}$ are shown in Fig. 10. The k_q

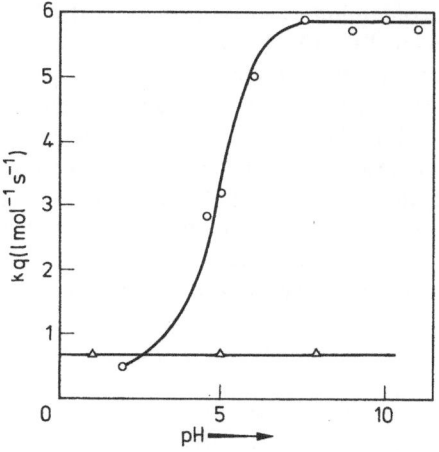

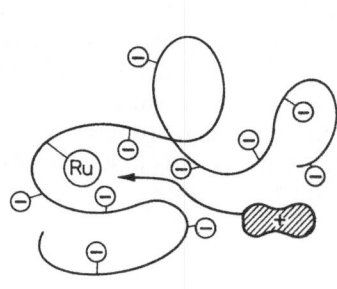

Fig. 10. pH dependences of k_q for (○) Ru[P(AA-Vbpy)](bpy)$_2^{2+}$ and (△) Ru (bpy)$_3^{2+}$. Ru; 10 μM, at 30 °C Measured at the MV^{2+} concentration below 1 mM

Fig. 11. Electrostatic attraction of a cationic substrate by a polyanion

of Ru(bpy)$_3^{2+}$ was independent of pH, while that of *9* increased sharply around pH 5 with the increase of pH. The k_q at neutral and alkaline conditions for the polymer complex is 9 times as high as that of Ru(bpy)$_3^{2+}$. Since the pH, where the rate sharply changes, almost agrees with the pK_a of the polymer pendant carboxyl group, its dissociation into the carboxylate anion must cause the higher rate for the polymer complex. The anionic domain provided by the carboxylates attracts the cationic subsubstrate (MV^{2+}), making the apparent rate constant high (Fig. 11).

10

$(m = 0.14, n = 0.86, x = 0.096)$

The polymer pendant $Ru(bpy)_3^{2+}$ (10) [34a] containing carboxyl groups at the bpy ring showed similar pH dependences of k_q as in Fig. 10 [34b] when MV^{2+} was used as acceptor. The rate of electron transfer from the excited state of 10 to Cu^{2+} was about 9 times higher than from $Ru(bpy)_3^{2+}$, because the free bpy pendant ligand co-ordinates the Cu^{2+}, thus facilitating the interaction of the coordinated Cu^{2+} with the adjacent pendant Ru complex. The poly(p-styrenesulfonate) (PSS) enhanced the rate about ten times. The anionically charged PSS and the cationic polymer complex (10) form a polyion complex, which contains an anionic domain due to the dissociated PSS. The anionic domain attracts Cu^{2+} and enhances the rate.

$Ru(bpy)_3^{2+}$ was covalently linked to viologen units to give a model of photoreaction center (11–13) [35].

11 n = 2, 2 $C_3V^{2\oplus}C_2$—$Ru^{2\oplus}$
12 n =15, 2 $C_{16}V^{2\oplus}C_2$—$Ru^{2\oplus}$

13 PS—$V^{2\oplus}$—$Ru^{2\oplus}$

The emission from $Ru(bpy)_3^{2+*}$ of these compounds was extremely weak because of the rapid quenching by the adjacent viologen unit. The systems of mixed micells composed of *12*, $2C_{12}V^{2+}$ (*14*) and CTAC, and polysoap composed of *12* and PS-V^{2+}-C_{16} (*15*)

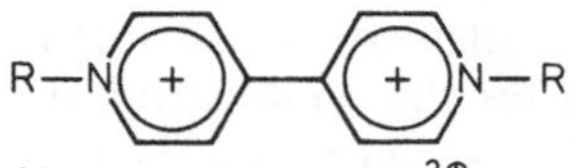

14 R = $C_{12}H_{25}$, 2 $C_{12}V^{2\oplus}$

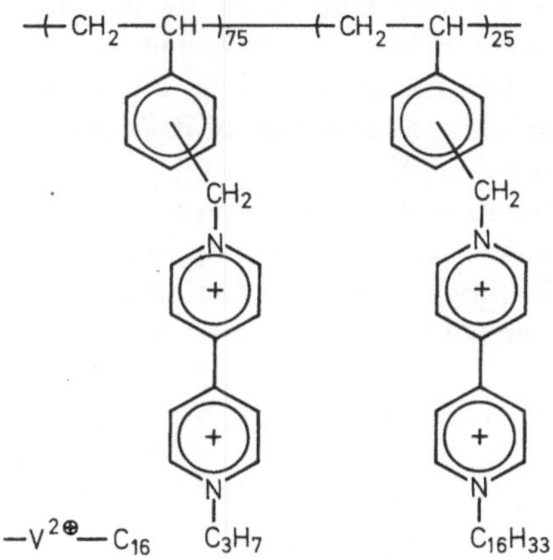

15 PS–$V^{2\oplus}$–C_{16}

showed high charge separation efficiencies to accumulate viologen cation radical in the presence of EDTA as sacrificial reducing agent. In these systems of mixed micells, the viologens containing long alkyl chains (*14*) or polymeric structure (*15*) are incorporated close to the Ru complex containing viologen groups (e.g., 12) as shown in Scheme 2. The mechanism of electron migration through the neighboring viologen units was proposed to explain the high efficiencies.

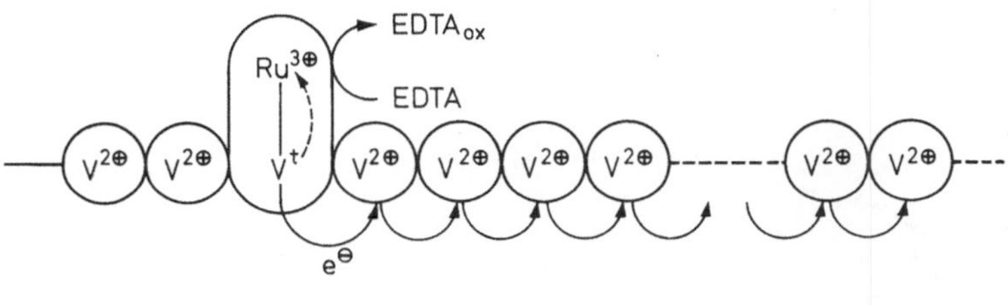

Scheme 2

Polymer complexes of $Ru(bpy)_3^{2+}$ (*16*, *17*) were prepared from poly(6-vinyl-2,2'-bipyridine) and poly(4-methyl-4'-vinyl-2,2'-bipyridine) [36]; the sensitized reduction of MV^{2+} with these complexes was studied by flash photolysis [37]. In the electron transfer

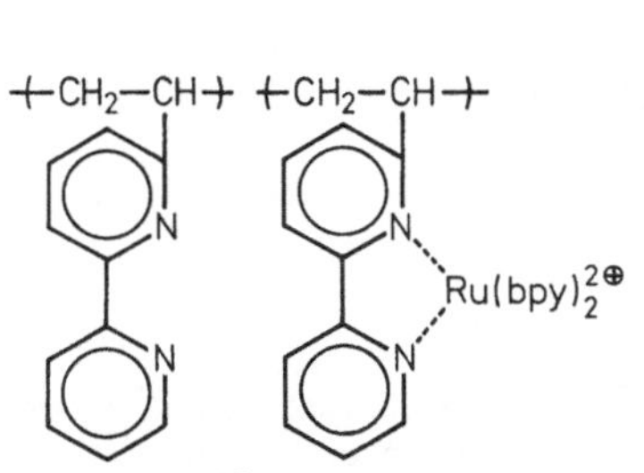

16 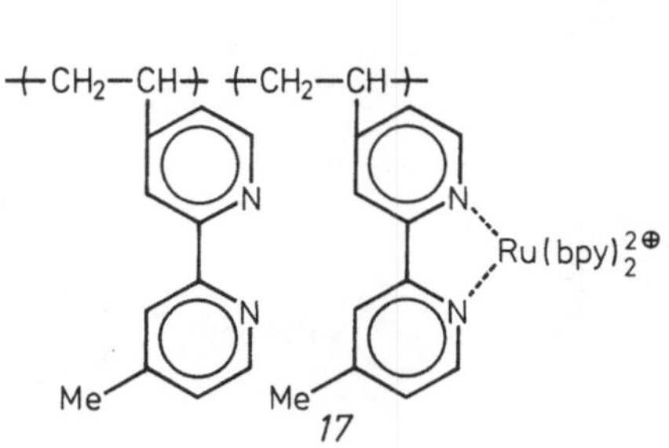   *17*

from the excited state of the polymer complex to MV^{2+}, smaller k_q values were observed than for the monomeric complex. The lower efficiencies for the polymer complex was considered to be due to their small diffusion coefficients. The different behaviours of *8* and *17* might be explained by the different backbone structure, molecular weight, and solvent used for the photochemical reaction. The geminate recombination of $Ru(bpy)_3^{3+}$ and $MV^{\ddagger}$ was faster for the polymer complex than $Ru(bpy)_3^{2+}$.

The ionic domain of polyelectrolytes also affects the photoinduced charge separation between the coexisting low molecular charged compounds [38]. The quenching of $Ru(bpy)_3^{2+*}$ by a zwitterion compound, dibenzylsulphonates viologen (*18*, BSV), was enhanced about 4 times by poly(vinylsulphate) (PVS) [39], and the back reaction

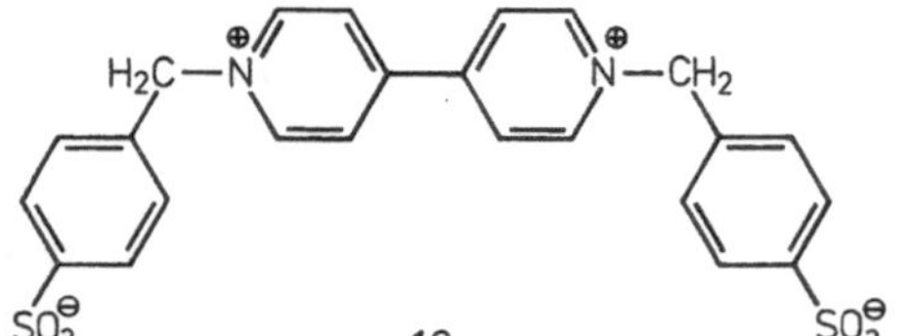

18

of the products ($Ru(bpy)_3^{3+}$ and $BSV^{\dagger}$) was retarded by one order of magnitude in the presence of PVS. The concentration of both the species on the anionic polymer chain as shown in Fig. 12 was postulated to be the reason for the high quenching rate. The retardation of the back reaction is due to the repulsion of $BSV^{\dagger}$ by the negative polyelectrolyte.

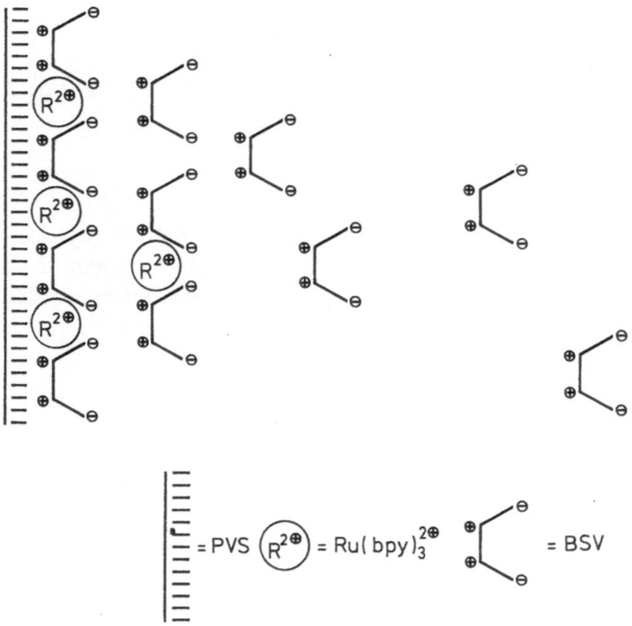

= PVS $\left(R^{2\oplus}\right)$ = Ru(bpy)$_3^{2\oplus}$ = BSV

Fig. 12. Electrostatic concentration of both the cationic reactants on a polyanion

As described in the first part of this section, MV⁺ can *reduce protons to give H₂* with a platinum catalyst. The presence of Pt colloid in the photoreaction mixture of $Ru(bpy)_3^{2+}$ polymer complex (derived from water soluble homopolymer of Vbpy), MV^{2+} and EDTA gave H_2 gas at almost the same rate as the mixture containing $Ru(bpy)_3^{2+}$ instead of polymer complex [40] (see Scheme 1). The turnover number of the Ru polymer complex exceeded 25 in 1 h's irradiation. The water insoluble polymer complex (*8*) showed almost the same activity when used as suspensions in a mixture of MeOH/H₂O = 1/1.

Oxidation of water to evolve O₂ is an important reaction for water photolysis as described in Sect. 2. Oxygen evolution from water by $Ru(bpy)_3^{3+}$ oxidation with RuO_2 catalyst was studied [41]. The authors established an electron relay system for water oxidation (Scheme 3) by the polymer $Ru(bpy)_3^{2+}$ complex and RuO_2 catalyst with PbO_2 as an oxidant [40].

In this system RuO_2 powders were coated with the water insoluble polymer complex (*8*) and used as suspensions. RuO_2 colloids were stabilized by the water soluble Ru complex prepared from poly(Vbpy) and used (see also Sect. 3.4).

$$\tfrac{1}{2}H_2O \Big)\Big(\frac{RuO_2^{\oplus}}{RuO_2} \Big)\Big(\begin{array}{c} Ru(bpy)_3^{2\oplus} \\ Polymer \\ complex \\ Ru(bpy)_3^{3\oplus} \end{array} \Big)\Big(\begin{array}{c} \tfrac{1}{2}Pb^{4\oplus} \\ \\ \tfrac{1}{4}Pb^{2\oplus} \end{array} \qquad \text{Scheme 3}$$

$$\tfrac{1}{4}O_2 + H^{\oplus}$$

Polystyrene pendant viologen (*19*) was prepared in order to use it as a polymeric electron mediator [42a]. The polymer itself can stabilize Pt colloids. The combined

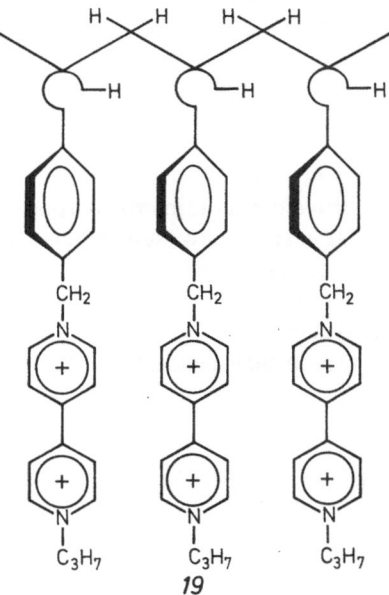

19

polymeric substance containing mediator (viologen) and Pt colloid catalyst generated H_2 under irradiation in the presence of $Ru(bpy)_3^{2+}$ and EDTA in aqueous solution. Visible spectrum of the cation radical of the pendant viologen showed an evidence of dimeric cation radical formation. ESR spectrum suggested electron exchange between the cation radicals. The electron migration among the pendant viologen groups was proposed [42a]. Although the efficiency of H_2 formation was lower for the present polymer system than that of monomeric viologen and Pt colloids stabilized by poly(vinylpyrrolidone), this is an interesting example to molecularly combine electron mediator with H_2-evolving catalyst.

A copolymer of styrene and acrylamide containing pendant viologen groups was used to stabilize Pt colloids. [42b] Photochemical formation of H_2 was observed with this combined mediator-catalyst system in the presence of $Ru(bpy)_3^{2+}$ and triethanolamine in water. The rate was, however, lower than the corresponding monomeric system. When $Na_2S_2O_4$ was used as reducing agent, the rate of chemical formation of H_2 was higher for the polymer system. Favorable electron transfer of $S_2O_4^{2-} \rightarrow$ pendant viologen $\rightarrow$ Pt colloids in the polymer domain was proposed.

Polymer pendant viologen was studied for the quenching efficiencies of the $Ru(bpy)_3^{2+*}$, the back reaction following quenching, the viologen radical yield and the effectiveness in H_2 production [43]. Viologen polymer was unfavorable for all these steps in comparison with monomeric MV^{2+}. A design of microenvironment around the pendant viologen groups was suggested to enhance the efficiency by ejecting the product.

The photoreduction of polymer pendant viologen by 2-propanol was reported to proceed by the successive two-electron transfer processes between the adjacent viologen units and the propanol which is a two-electron reducing agent [44]. Preferential formation of a dimeric cation radical of viologen observed was ascribed to the polymeric structure and the two-electron process. These fundamental studies on polymeric electron mediators contribute to the construction of solar energy conversion systems.

3.3 Polymer Systems as Solid Phase

As described above, polymeric materials provide specific microenvironment in solution which contributes much to construct solar energy conversion systems. Macroheterogeneous systems constructed from polymers are of great value especially from the practical point of view.

Photochemical reactions in solid polymers were studied to find that MV^{2+} is reduced photochemically by carbohydrates in their solid phase [45] (Scheme 4).

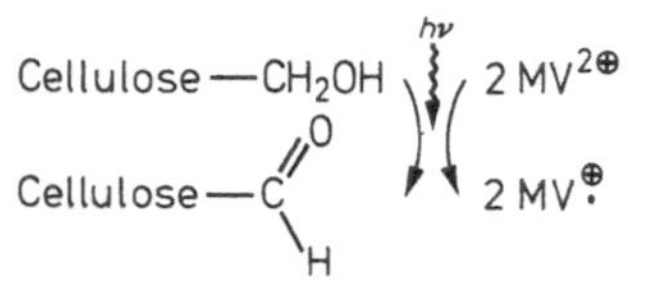

Scheme 4

For example, a cellulose paper adsorbing MV^{2+} turned deep blue under irradiation due to the accumulation of MV^{+}. Thus, the photochemical reaction occurred in the solid phase, whereas water hindered the reaction remarkably [46]. The interaction between the MV^{2+} and the reducing groups in a solid state is important for the photoreaction. The reaction is regarded as the photoinduced transfer of the reducing power of carbohydrate to MV^{2+}. The solid phase of gelatine [47] and synthetic polymers such as poly(vinylalcohol), poly(acrylic acid) and poly(vinylpyrrolidone) also brought about photoreduction of MV^{2+}.

Table 2. Effects of $Ru(bpy)_3^{2+}$ and Na_2EDTA on the photoreduction of MV^{2+} in cellulose solid phase: irradiated with 470 nm monochromatic light *in vacuo* at 30 °C

Components adsorbed on cellulose			Initial rate of MV^+ accumulation $(10^{-8}\ mol \cdot cm^{-2} \cdot min^{-1})$
—	—	MV^{2+}	0.011
—	$Ru(bpy)_3^{2+}$	MV^{2+}	0.012
Na_2EDTA	$Ru(bpy)_3^{2+}$	MV^{2+}	0.385

Polymer solids also work as a carrier of photochemical reaction components. The irradiation of a cellulose paper after adsorbing EDTA, $Ru(bpy)_3^{2+}$ and MV^{2+} induced rapid formation of MV^+ in the solid phase (Table 2) [45, 48]. The quenching experiments showed that a photoinduced electron relay of EDTA $\xrightarrow{e^-}$ $Ru(bpy)_3^{2+}$ $\xrightarrow{e^-}$ MV^{2+} occurs in the solid phase just like in the solution. In this reaction the main path for the MV^+ formation is through $Ru(bpy)_3^{2+}*$, and the rate of direct reduction of MV^{2+} by cellulose molecule is very small. Such an electron relay occurred also in a gelatine film [47].

The reducing power of the MV^+ formed in a solid phase can be transferred to liquid phase. $Ru(bpy)_3^{2+}$ and MV^{2+} were adsorbed in water-swollen chelate resin beads containing iminodiacetic acid (IDA) groups. The irradiation of the beads induced rapid formation of MV^+ in the solid phase through the electron relay of IDA $\xrightarrow{e^-}$ $Ru(bpy)_3^{2+}$ $\xrightarrow{e^-}$ MV^{2+} [49]. When the beads were irradiated in water contain-

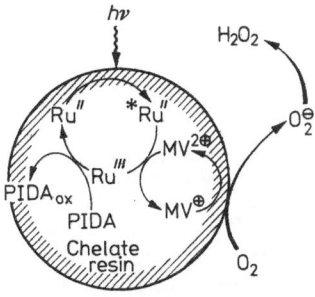

Bulk solution

Fig. 13. Generation of hydrogen peroxide in $Ru(bpy)_3^{2+}$-MV^{2+}/PIDA resin-O_2-containing aqueous solution

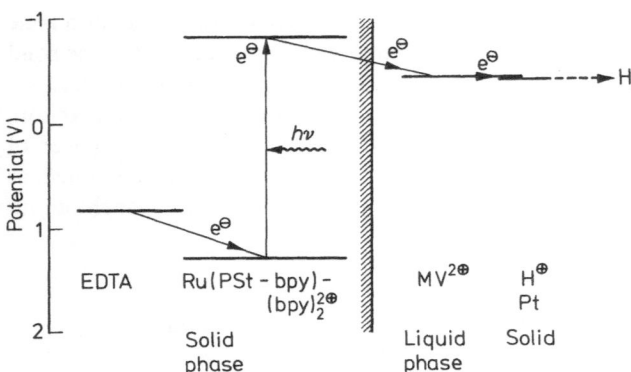

Fig. 14. Photoinduced charge separation at the solid-liquid interface followed by hydrogen evolution in the liquid phase

ing dissolved oxygen, hydrogen peroxide was formed by interfacial electron transfer from $MV^{\pm}$ formed in the solid phase to O_2 in the liquid phase (Fig. 13). The rate of H_2O_2 formation reached ca. $8\,\mu M\,min^{-1}$ for the $0.2\,g$ resin adsorbing $150\,\mu mol$ $Ru(bpy)_3^{2+}/g$ resin and $50\,\mu mol\,MV^{2+}/g$ resin suspended in O_2 saturated water (50 ml) at pH 7.4.

A photoinduced electron relay system at solid-liquid interface is constructed also by utilizing polymer pendant $Ru(bpy)_3^{2+}$. The irradiation of a mixture of EDTA and water-insoluble polymer complex $(Ru(PSt\text{-}bpy)(bpy)_2^{2+}$, prepared by Eq. (15)) deposited as solid phase in methanol containing MV^{2+} induced $MV^{\pm}$ formation in the liquid phase [9]. The rate of $MV^{\pm}$ formation was $4\,\mu M\,min^{-1}$. As shown in Fig. 14, photoinduced electron transfer occurs from EDTA in the solid to MV^{2+} in the liquid via $Ru(bpy)_3^{2+}*$. The protons and Pt catalyst in the liquid phase brought about H_2 evolution. One hour's irradiation of the system gave $9.32\,\mu l\,H_2$ after standing 12 h and the turnover number of the Ru complex was 7.6 under this condition. The apparent rate constant of the electron transfer from $Ru(bpy)_3^{2+}$ in the solid phase to MV^{2+} in the liquid was estimated to be higher than that of the entire solution system. The photochemical reduction and oxidation products, i.e., H_2 and $EDTA_{ox}$ were thus formed separately in different phases. Photoinduced electron relay did not occur in the system where a film of polymer pendant Ru complex separates two aqueous phases of EDTA and MV^{2+} [9] (see Fig. 15c).

Crosslinked poly(styrene) beads (20) were prepared with contain $Ru(bpy)_3^{2+}$ anchored at oligo(oxyethylene) spacer group [50]. Vigorous stirring was needed for the suspended complex (20) to sensitize the photoreduction of MV^{2+} in the presence of

20

Ru–EDTA (bpy)$_3^{2\oplus}$ MV$^{2\oplus}$	Occurrence of the photoinduced electron relay EDTA $\xrightarrow{e^{\ominus}}$ Ru(bpy)$_3^{2\oplus}$ $\xrightarrow{e^{\ominus}}$ MV$^{2\oplus}$	Ref.
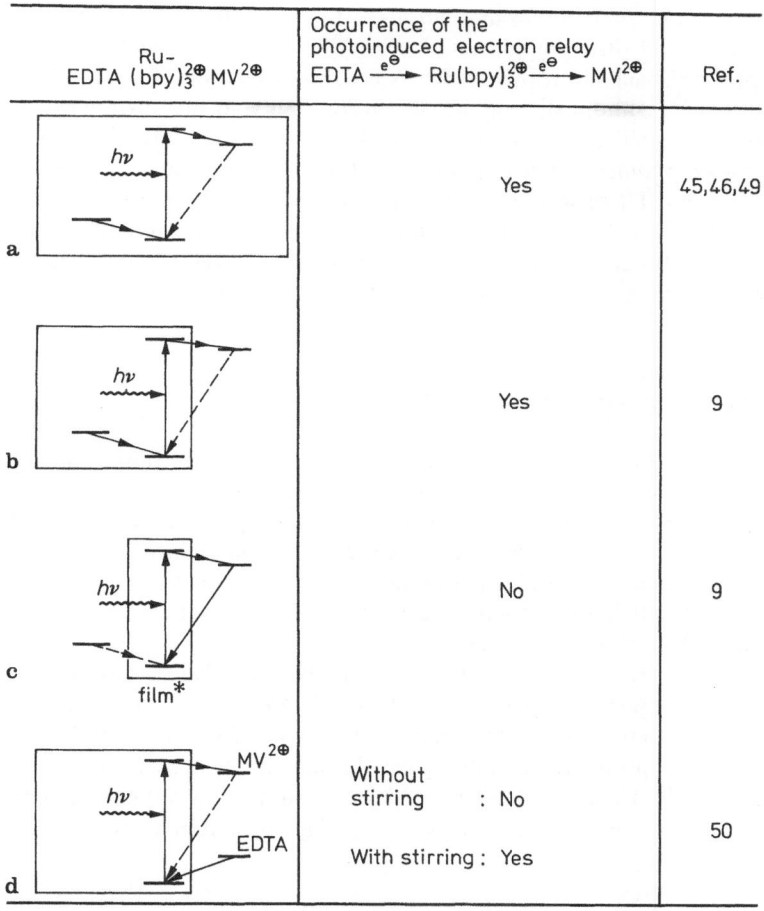 a	Yes	45,46,49
b	Yes	9
c film*	No	9
d MV$^{2\oplus}$ EDTA	Without stirring : No With stirring : Yes	50

Fig. 15. a–d. Photoinduced electron relay in the solid phase and at the solid-liquid interface. □: Solid phase (outside is the liquid phase); →: Electron transfer; *: The film separates two liquid phases

EDTA in water. Ru(bpy)$_3^{2+}$ and its dicinnamate were immobilized in a membrane prepared from cinnamate of poly(vinyl alcohol) by photocrosslinking [51], and the immobilized complex sensitized the photoreduction of MV^{2+}.

The photoinduced electron relay systems in the solid phase and at the solid-liquid interface containing Ru(bpy)$_3^{2+}$ in the solid phase are summarized in Fig. 15. The electron transfer from Ru(bpy)$_3^{2+}$* to MV^{2+} is very facile and occurs even at the solid-liquid interface. The back electron transfer of the products (Ru(bpy)$_3^{2+}$ + MV$^{+\cdot}$ → Ru(bpy)$_3^{2+}$ + MV^{2+}) is so rapid, however, that MV$^{+\cdot}$ can be accumulated only when Ru(bpy)$_3^{3+}$ is rapidly reduced by a reducing agent such as EDTA. If the reduction of Ru(bpy)$_3^{3+}$ by EDTA shall compete with that by MV$^{+\cdot}$, both *EDTA and Ru complex*, or at least both *EDTA and MV^{2+} should exist in the phase*. The systems (a) and (b) of Fig. 15 reduce Ru(bpy)$_3^{3+}$ by EDTA with the Ru complex in the same phase, leading to the accumulation of $MV^{\cdot+}$. In the system (c), with $Ru(bpy)_3^{2+}$ and EDTA in different phases, $Ru(bpy)_3^{3+}$ is not reduced by EDTA, resulting in the prefered back electron transfer. Mere recombination of the products prevails in the system (d) without stirring the suspension, but vigorous stirring causes $MV^{\cdot+}$ accumulation by repelling $MV^{\cdot+}$ and allowing the access of EDTA to $Ru(bpy)_3^{3+}$.

Macroheterogeneous photoconversion systems can thus be constructed by utilizing solid phase of polymers. These results on the solid phase and solid-liquid interface electron relay are surely of use when designing macroheterogeneous conversion systems. The findings obtained in these studies were used to develop a new photo-diode composed of polymer pendant $Ru(bpy)_3^{2+}$ film as described in Section 4 of Chapter 5.

3.4 Polymer Supported Metal Colloids as Catalyst

As described in Section 3 of Chapter 2, multi-electron processes are important for designing conversion systems. Noble metals are most potent catalysts to realize a multi-electron catalytic reaction. It is well known that the activity of a metal catalyst increases remarkably in a colloidal dispersion. *Synthetic polymers* have often been used to *stabilize the colloids*. Colloidal platinum supported on synthetic polymers is attracting notice in the field of photochemical solar energy conversion, because it reduces protons by $MV^{\cdot+}$ to evolve H_2 gas. [1a]

In the photochemical H_2 generation system composed of EDTA, $Ru(bpy)_3^{2+}$, MV^{2+}, and Pt colloid supported on PVA (see Scheme 1), smaller Pt particles gave a higher activity for H_2 production [52]. The same system using PVA and poly(vinyl-pyrrolidone) (PVPyr) as protective colloids gave the maximum activity at the Pt particle size of 30 Å [53]. Usually, smaller particles should show higher activity because of the larger surface of the catalyst. Since H_2 formation from protons

Table 3. Light-Induced Hydrogen Production by Platinum Catalysts Prepared in the Presence of Various Polymers

Polymer		Catalyst		Hydrogen production
name (D.P.)	protective value	form	av. diameter (Å)	(l/day/l)[a]
Polyethyleneimine	0.04	complex	—	0.1
Polyacrylamide	1.3	ppt	—	—
Poly(vinyl methyl ether)	—	colloid[b]	—	0.1
Poly(vinyl alcohol) (500)	5	colloid	34	2.1
Polyacrylic acid	0.07	colloid	29	2.5
Hydroxyethyl Cellulose	—	colloid	29	3.1
Polyvinylpyrrolidone (3250)	50	colloid	29	4.0
Gum Arabic	—	colloid	20	3.3
Gelatine	90	colloid	16	3.2

[a] $Ru(bpy)_3^{2+} = 5 \times 10^{-5} M, MV^{2+} = 5 \times 10^{-3}$ M, EDTA $\cdot$ 2Na $= 5 \times 10^{-2}$ M, Pt $= 6.6 \times 10^{-5}$ M;
[b] Platinum metal was precipitated after irradiation

is a two-electron process, two molecules of $MV^{\ddagger}$ must donate electrons to protons at the same time or successively in a short time. When the Pt particle size is too small, the contact of two molecules of $MV^{\ddagger}$ would be hindered; this must make the two-electron process difficult.

The protective ability of a polymer to stabilize colloids is represented by protective value. It is the weight (g) of gold that is stabilized by 1 g protecting polymer on the addition of 1% NaCl aqueous solution [54]. The Pt colloids on various polymers and their catalytic activities for H_2 production in EDTA-Ru(bpy)$_3^{2+}$-MV^{2+}-Pt colloids/polymer system are shown in Table 3 [55]. PVPyr, gum arabic, gelatine, and hydroxyethyl cellulose showed especially high activity. It was concluded that the protection against coagulation of the particles is more important than the formation of small particles. Optimal concentration of Pt colloids was observed for the H_2 generation [56], because Pt colloids also catalyze the hydrogenation of MV^{2+} which then becomes inactive as electron mediator.

Immobilization of the colloidal catalyst on resins was studied in order to use them repeatedly as stable solid catalyst. The citric acid reduction of poly(hydroxyethyl-methacrylate) containing H_2PtCl_4 yielded immobilized Pt colloids which are dispersed in that polymer [57]. It could be used as H_2 generating catalyst repeatedly without activity loss in the system of EDTA-Ru(bpy)$_3^{2+}$-MV^{2+}.

Colloidal catalysts evolve oxygen gas according to Scheme 5.

$$\begin{array}{c} 1/2\,H_2O \\ 1/4\,O_2 + H^{\oplus} \end{array} \bigg) \bigg(\begin{array}{c} RuO_2^{\oplus} \\ RuO_2 \end{array} \bigg) \bigg(\begin{array}{c} Ru(bpy)_3^{2\oplus} \\ Ru(bpy)_3^{3\oplus} \end{array} \bigg) \bigg(\begin{array}{c} \text{Oxidant} \\ \text{Reduction} \\ \text{product} \end{array} \qquad \text{Scheme 5}$$

Polymer pendant Ru(bpy)$_3^{2+}$ prepared from homopolymer of Vbpy formed stable RuO_2 colloid from RuO_4 in an aqueous solution [40]. The colloid generated O_2 in the presence of PbO_2 as an oxidant according to the Scheme 5. Colloids of $RuO_x \cdot IrO_x$ on Y-zeolite were formed by treating $RuCl_3$ and K_3IrCl_6 at 200 °C. They showed a high activity for O_2 evolution under irradiation in Scheme 5 when $Co(NH_3)_5Cl^{2+}$ was used as an oxidant [58].

These metal and metal oxide catalysts must work as a kind of electron pool which brings about multi-electron process for H_2 and O_2 generation. Silver colloids were studied as electron pool for H_2 formation under γ-irradiation in the aqueous system composed of Ag° colloids, acetone, 2-propanol and SDS [59]. The colloids (average diameter; 140 Å) of 2.5×10^{-4} M can store 1 coulomb/l, corresponding to the storage of 450 electrons/particle [60].

A colloidal polynuclear metal complex was proved to be an effective catalyst for water photolysis in combination with Ru(bpy)$_3^{2+}$ probably because its capability of multi-electron process. The details are described in the next chapter.

4 Water Photolysis by Polymer Systems

Polymers play important roles in water photolysis. For multi-electron processes, *polymer supported metal colloids* or *colloidal polynuclear metal complexes* are very useful as *catalysts*. Unstable semiconductors with a small bandgap which photolyse water under visible irradiation can be stabilized by coating with a conductive polymer film. Hydrogen and oxygen producing water photolysis by visible irradiation is described in this chapter.

4.1 Polymer Supported Metal Colloids

The combination of Scheme 1 and 5 with removing sacrificial reductant and oxidant should lead to a water photolysis system as represented by Fig. 16. The thermodynamics of the reactions is suitable to photolyse water into H_2 and O_2. This system is very similar to the conversion model of Fig. 3 with one difference that the electron mediator (T_2) lacks in the system of Fig. 16. The irraditation of an aqueous system containing $Ru(bpy)_3^{2+}$, MV^{2+}, Pt colloid supported on PVA and RuO_2 sol supported on poly(styrene-co-maleic anhydride) induced water photolysis to give H_2 and O_2 simultaneously [61] according to the Fig. 16. Although water photolysis by the system of Fig. 16 is thermodynamically possible, the lack of O_2 analysis data [61] and the poor reproducibility of the complex reaction are unsolved problems. The reduction of MV^{2+} under visible irradiation with poly(styrene-co-maleic anhydride) [62], which was used as protective colloid for the RuO_2 in that paper [61], produces H_2 at the Pt catalyst. Generally, the possibilities of MV^{2+} photoreduction by various organic substances should be considered in order to avoid the misunderstanding what is actually happening in a photochemical reaction system containing MV^{2+}. The system composed of a suspension of TiO_2 doped with Pt as well as RuO_2, $Ru(bpy)_3^{2+}$, and MV^{2+} claiming water photolysis by visible irradiation [63] seems to involve the similar problems as the above system.

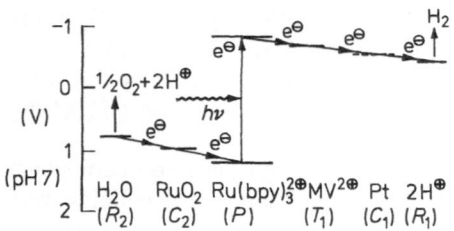

Fig. 16. A model of water photolysis

4.2 Polynuclear Metal Complexes

A polynuclear metal complex is a promising catalyst for multi-electron processes due to the polymeric structure of the metal ions. The authors found that a simple aqueous system of prussian blue (PB) and $Ru(bpy)_3^{2+}$ decomposes water under visible irradiation to give H_2 and O_2 simultaneously [64]. The PB is a polynuclear iron complex of the composition $KFe^{III}[Fe^{II}(CN)_6]$ forming a colloidal suspension in water [65] (Fig. 17). The evolved H_2 and O_2 were analyzed both by gas chromatogram and mass spectrum. The photolysis of water containing D_2O and $H_2^{18}O$ showed that both of the gases were generated by water decomposition. The addition of KCl accelerated the photolysis to show an optimal concentration of the salt. Since the coagulation of the PB colloids

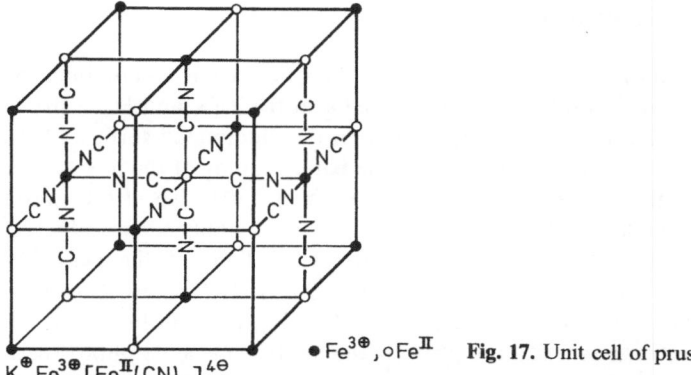

$K^{\oplus}Fe^{3\oplus}[Fe^{II}(CN)_6]^{4\ominus}$ $\bullet\,Fe^{3\oplus}, \circ\,Fe^{II}$ **Fig. 17.** Unit cell of prussian blue

prevented the photolysis, the colloidal state is important for the catalysis. Both the PB (λ_{max} = 700 nm) and the $Ru(bpy)_3^{2+}$ (λ_{max} = 452 nm) must be irradiated for water photolysis and are proved to act as catalyst. Experiments showed that PB not only reduces protons to H_2 under irradiation, but also oxidize water to O_2 with $Ru(bpy)_3^{3+}$ as oxidant. It was shown that PB quenched $Ru(bpy)_3^{2+}*$ partly by oxidative electron transfer to give $PB \cdot e^-$ and $Ru(bpy)_3^{3+}$. These results have led to the mechanism as shown in Eqs. (18) to (23).

$$Ru(bpy)_3^{2+} \xrightarrow{452\,nm} Ru(bpy)_3^{2+*} \tag{18}$$

$$Ru(bpy)_3^{2+*} + PB \rightarrow Ru(bpy)_3^{3+} + PB \cdot e^- \tag{19}$$

$$Ru(bpy)_3^{3+} + PB \cdot e^- \rightarrow Ru(bpy)_3^{2+} + PB \tag{20}$$

$$Ru(bpy)_3^{2+*} + PB \cdot e^- \rightarrow Ru(bpy)_3^{3+} + PB \cdot 2e^- \tag{21}$$

$$PB \cdot 2e^- + 2H^+ \xrightarrow{700\,nm} PB + H_2 \tag{22}$$

$$2\,Ru(bpy)_3^{3+} + H_2O \xrightarrow{PB} 2\,Ru(bpy)_3^{2+} + 1/2\,O_2 + 2H^+ \tag{23}$$

The polyneclear and mixed valent structure of the PB allow multi-electron processes of Eqs. (22) and (23) as well as both the reduction and oxidation catalyses. The reactions are schematically shown in Fig. 18.

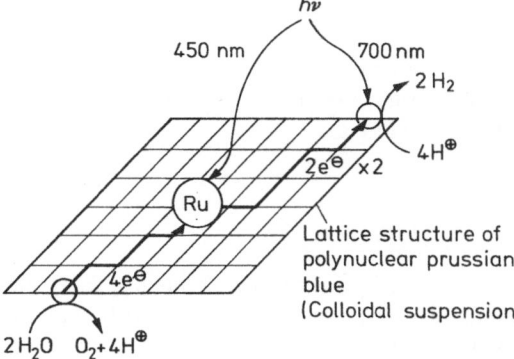

Fig. 18. Water photolysis with prussian blue and $Ru(bpy)_3^{2+}$

4.3 Polymer Coated Semiconductor

The UV irradiation of an n-type titanium dioxide (TiO_2) electrode in combination
with a Pt counter electrode photolyses water to give O_2 at the TiO_2 and H_2 at the Pt
electrodes [66]. The photoinduced charge separation on the illuminated surface of the
so-called liquid junction semiconductor is due to the band bending at the interface as
shown in Fig. 19. Another n-type semiconductor such as ZnO also effects water
photolysis. A problem in this system is that the bandgap of stable semiconductors
such as TiO_2 (3.0 eV) or ZnO (3.2 eV) allows to utilize light of the near UV
region only. The n-type semiconductor such as Si (1.1 eV), GaAs (1.35 eV), and CdS
(2.4 eV) are all unstable in water under irradiation. A poly(pyrrole) film coated on
an unstable n-Si by electropolymerization protects the electrode against photocorrosion [67a] and generates a stable photocurrent as described in the next chapter. The
poly(pyrrole) film also prevented n-CdS from photocorrosion. Such a stabilized narrow
bandgap n-type semiconductors can be used for water photolysis in combination
with Pt counter electrode. Visible light irradiation of a CdS'electrode coated with
poly(pyrrole) incorporated with RuO_2 dispersions gave O_2 on the surface, and H_2
on the Pt counter electrode [67b] (Fig. 19). This modified CdS electrode allowed 68 %
of the photogenerated holes to oxidize water for O_2 production, and 32 % to corrode
CdS.

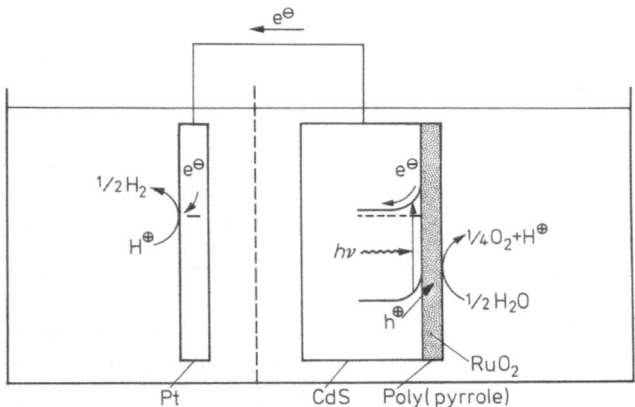

Fig. 19. Water photolysis by polymer coated semiconductor of narrow bandgap

Since photoinduced charge separation occurs easily at the *semiconductor*-water
interface, its *stabilization with polymer coating* is one of the most promising approaches
to water photolysis by visible irradiation.

5 Solar Cells

The efficiency of the amorphous silicon solar cell has reached about 8 %, and the
cost of the electricity generated is estimated to become comparable with that by
fossil fuels in near future. In spite of this situation, the production of a light-weighing

and cheap solar cell from organic materials such as polymers simply by molding is attracting many researchers. A polymer film with the necessary function is a promising material for that purpose. In this chapter, solar cells composed of polymer films, polymer coatings to to stabilize liquid junction photovoltaic cells, photogalvanic cells composed of polymers, and a new photodiode composed of a polymer film containing a metal complex are described.

5.1 Polymer Films as Solar Cells

Recently, intense interest has been paid on doped poly(acetylene) because its film [68] showed markedly high conductivity on doping [69, 70, 71] and the n- and p-type conductivities are depending on the dopants. The confirmation of p-n junction formation with the p- and n-type $-(CH)_x$ has roused great expectations to produce a polymer film solar cell. [72a]

The doping of $-(CH)_x$ with electron acceptors or donors increases the electrical conductivity by 12 orders of magnitude. Doping with an acceptor (e.g., alkaline metal) or an donor (e.g., halogen, AsF_5) gives a p- or n-semiconductor, respectively [70, 72a]. The conductivity increases drastically with the dopant concentration, and reaches saturation at the concentration of 2 to 3 %. The conductivity [72b] of cis-$[CH(AsF_5)_{0.14}]_x$ at 300 K was 560 S cm^{-1}. The doping transfers the polymer almost to metalic conductivity of the order of 10^3 S cm^{-1}.

The doped poly(acetylene) forms various junctions such as a) a p-n junction from p- and n- $-(CH)_x$, b) a hetero-Schottky junction from the inorganic semiconductor and metalic $-(CH)_x$, and c) a heterojunction from the inorganic semiconductor and semiconducting $-(CH)_x$. The bandgaps of $-(CH)_x$ (trans-; 0.6 eV, cis-; 0.9 eV) efficiently absorb the solar irradiation.

A solar cell was constructed from a p-$-(CH)_x$ film and a single crystalline n-Si wafer. The current-potential characteristic of the battery is shown in Fig. 20 [73]. A conversion efficiency of 4.3 % was obtained. The most sincere problem for poly-(acetylene) is its instability against air oxidation. If it could be solved, practical use of light weight and easily moldable solar cell from polymer films might be possible.

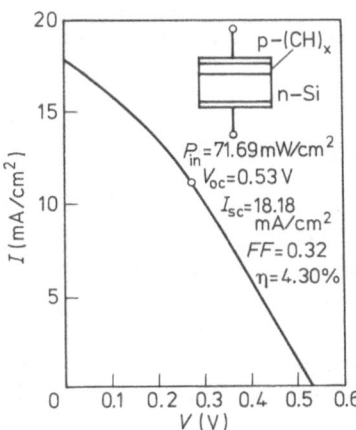

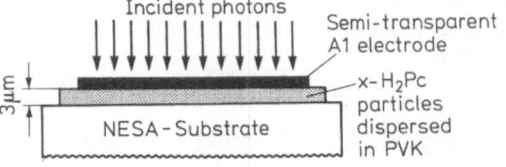

Fig. 21. Solar cell composed of a polymer film containing sensitizer

Fig. 20. Solar cell composed of poly(acetylene) film (Light is irradiated from the p-(CH)$_x$ side)

Sandwich type solar cells are composed of a thin layer of sensitizer such as phthalocyanine (Pc) put between metal electrodes [74, 75]. The dispersion of the sensitizer in synthetic polymers increases the conversion efficiency [76]. The photovoltaic cell composed of a 0.5–5 μm film of poly(vinylcarbazole) (PVK) dispersed with x-H_2Pc is shown in Fig. 21. The conversion efficiency of the cell: nesa glass[1]/x-H_2Pc, PVA/Al, reached a maximum value of 3.8 % at the H_2Pc concentration of ca. 60 wt % under 670 nm monochromatic light irradiation. Much higher efficiency of poly-(vinylidenefluoride) than PVK for the cell; In_2O_3/α-H_2Pc, PVDF/Al, suggested the effect of micro electric field of the polar PVDF [77]. The effect of electroconductive polycation on the same cell was studied [78]. A flexible photovoltaic cell was produced by depositing a merocyanine dye (21) layer on a transparent electro-conductive film which is polyester coated with indium-tin oxide (ITO) [79]. A

sunlight efficiency of 0.2 % was obtained for this Al/dye/ITO/polyester cell. The problem of this type of solar cell is the low conversion efficiency under white light irradiation, although the efficiency under monochromatic light is as high as 6 %.

A MIS (Metal-Insulator-Semiconductor) type cell: nesa glass/NiPc/poly(ethylene) (PE)/Al, was constructed, and its I–V characteristics were studied [80]. The insulating PE layer caused higher open circuit voltage (V_{oc}) of 0.65 V than the cell without PE (0.48 V). A film of the complex of poly(2-vinylpyridine) with I_2 formed a cell: SnO_2/P2VP-I_2/Pt, showing a conversion efficiency of 0.05 % under white light irradiation [81]. A thin film of poly(ethylene oxide) incorporated with a polysulfide redox couple was used as a plastic electrolyte in the photovoltaic cell of n-CdS/plastic electrolyte-Na_2S_4/p-CdTe [82].

Ferroelectric substances such as $LiNbO_3$ or $BaTiO_3$ were found to show anomalous photovoltaic effects (APV) of the order of 10^{3-5} V [83]. A thin film of a ferroelectric polymer such as PVDF also was found to show APV effects [84]. V_{oc} of 2.5×10^4 V and short circuit current (I_{sc}) of the order of nA/cm^2 were reported. Although the output is very small up to now, it could be noticed as a specific photoeffect of a thin polymer film.

5.2 Polymer Film Coating to Stabilize Liquid-junction Photovoltaic Cells

As described in Section 3 of Chapter 4, the stabilization of n-Si electrode by coating with poly(pyrrole) has attracted much attention. The *stabilization of a small bandgap n-semiconductor* electrode against oxidation is of great value not only to convert visible light into chemical energy, but also to construct liquid-junction solar cells operated under visible irradiation. The poly(pyrrole) film is usually electropolymerized on the semiconductor electrode dipped in the aqueous solution of pyrrole. The remarkable stabilizing effect of poly(pyrrole) film on polycrystalline n-Si is shown in Fig. 22 [67]. The photocurrent obtained under irradiation in the aqueous solution of

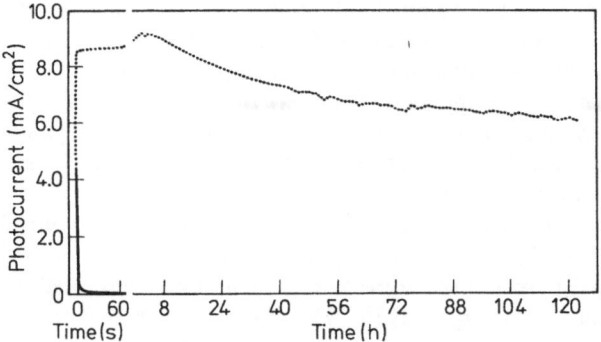

Fig. 22. Short-circuit photocurrent of naked (—) and polypyrrole covered (...) polycrystalline n-type Si electrode in aqueous 0.15 M $FeSO_4$, 0.15 M $FeNH_4(SO_4)_2 \cdot 12\,H_2O$ and 0.1 M Na_2SO_4 at pH 1 under tungsten-halogen illumination at 143 mW/cm^2. Air was not excluded.

Fe^{2+}/Fe^{3+} at the naked n-Si electrode drops to zero in less than 30 s, whereas the poly(pyrrole)coated n-Si gives a stable photocurrent for 5 days. The power conversion efficiency was 3 % corresponding to I_{sc} of 3.2 mA/cm^2, V_{oc} of 0.39 V, and a fill factor of

0.6 under tungsten illumination. The protection by the polymer is attributed to the coverage of potential oxide forming sites on the electrode surface. The polymer layer efficiently transmits holes generated in the semiconductor away from the electrode-polymer interface to ferrous ions. The important characters of the poly(pyrrole) for that function are the solution permeable, conducting, and may by complex-forming properties. These characters must lead to efficient transfer of the holes to the Fe^{2+}/Fe^{3+} redox couples before the oxidation of the electrode takes place.

One problem for the coated system is that the film is peeled off after prolonged irradiation. In order to have a more adhesive film, the surface of n-Si was modified with N-(3-trimethoxysilylpropyl)pyrrole (22). Pyrrole was then electrodeposited on this modified electrode as shown in Eq. (24) [85]. The durability of the coated poly(pyrrole) was improved by such a treatment of n-Si surface. The n-Si electrode coated only with poly(pyrrole) gave a declined photocurrent from 6.5 to 1.8 mA cm^{-2} in less than 18 h, while the poly(pyrrole) coated n-Si treated at first with 22 as Eq. (24) gave a stable photocurrent of 7.6 mA cm^{-2} for 25 h. When an n-Si electrode was coated with Pt layer before the deposition of poly(pyrrole), the stability of the semiconductor was improved remarkably (ca. 19 days) [85b]. A power conversion efficiency of 5.5 % was obtained with iodide/iodine redox electrolytes.

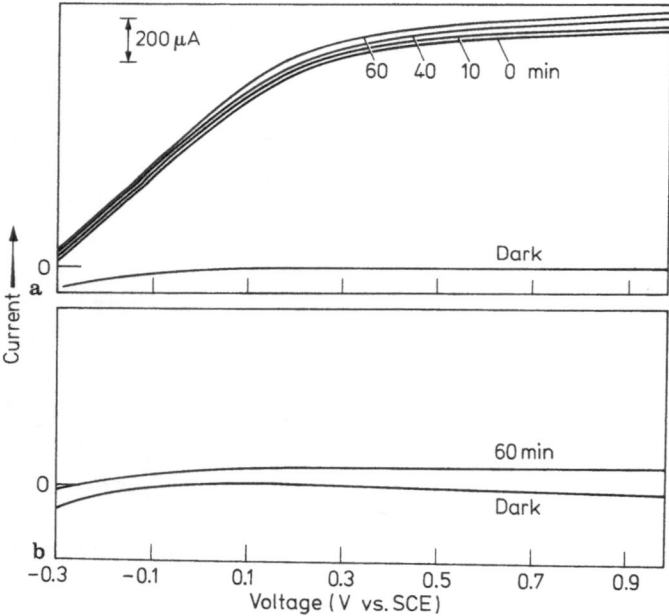

Fig. 23. Time dependence of steady-state I–V behaviour (scan rate: 100 mV/s) for (a) polymer-coated n-GaAs electrode (area: ~0.1 cm^2) and (b) bare n-GaAs photoanode in contact with the I_3^-/I^- (0.5/0.5 M) redox electrolyte (pH = ~5). The illuminated electrode (light intensity 53 mW/cm^2) was maintained at approximately short-circuit condition (0.3 V vs. SCE) for the duration shown, after which the potential scans were initiated. The initial level of the I–V curve for the bare electrode was close to that seen at 0 min for the coated sample. The electrolyte was stirred in all cases

An n-GaAs electrode was also stabilized by poly(pyrrole) coating [86]. A film of polymer pendant $Ru(bpy)_3^{2+}$ like 8 also stabilized n-GaAs against photocorrosion [87]. The polymer film of 8 was casted on the electrode from its DMF solution. The current-voltage (I–V) behaviours of the coated and bare electrodes in I_3^-/I^- redox aqueous solution in the dark and under illumination are shown in Fig. 23. The coated electrode shows remarkable stabilizing effect under illuminated condition. The performance efficiency of the coated n-GaAs in 12% at V_{oc} of 1.09 V, I_{sc} of 10.89 mA/cm^2, and fill factor of 0.71. The role of the Ru complex in the charge transport across the polymer layer was discussed.

If small bandgap semiconductors could be stabilized in water for much more prolonged time by polymer coating, it must lead to developments of efficient photochemical cell and water photolysis system.

5.3 Photogalvanic Cells Composed of Polymers

When a photochemical process in solution gives a photoresponse at the electrode, the system can form a photogalvanic cell. The photoinduced redox reaction is typical for photogalvanic cells. The most well known photoredox system is thionine (TH^+; 23) and ferrous ion [88]. The excited TH^+ is reduced by Fe^{2+} to give TH_2^+ and Fe^{3+} (Eq. (25)).

$$\underset{\text{(violet)}}{TH^+} + Fe^{2+} + H^+ \underset{\text{dark}}{\overset{h\nu}{\rightleftharpoons}} \underset{\text{(colorless)}}{TH_2^+} + Fe^{3+} \qquad (25)$$

Since illumination shifts the equilibrium to the right, the aqueous violet solution is decolorized. When the light is cut off, the color turns again to violet due to the equilibrium shift to the left. The reversible photoredox mixture gave a photopotential when irradiated in a photochemical cell composed of light and dark

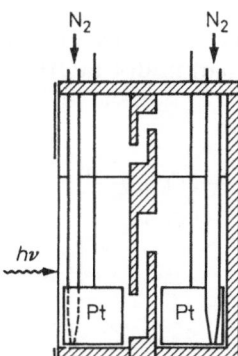

Fig. 24. Photogalvanic cell composed of light and dark chambers

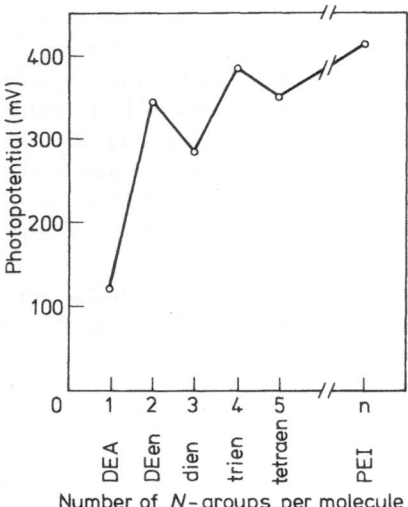

Fig. 25. Photopotentials of systems amine/thionine (TH) as function of the number of amino or imino groups (*N*-groups) per molecule of the amine. [TH] = 10^{-5} mol/l; [*N*-groups] = $5 \cdot 10^{-3}$ mol/l; pH 8. DEA; Diethylamine, DEen; N,N′-diethylethylene-diamine, dien; Diethylenetriamine, trien; Triethylenetetramine, tetraen; Tetraethylenepentamine, PEI; Poly(ethyleneimine)

chambers [11, 89)] (see Fig. 24). Tolusafranine and EDTA showed the highest photopotential of 844 mV [90)], although it is an irreversible system accompanying EDTA decomposition.

Aliphatic amines act as reducing agents for the photoreduction of phenazine or phenothiazine dyes. It was found that polyamine induces a much higher photopotential than monoamine [91)]. The value of the photopotential was strongly dependent upon the number of amino or imino groups in one molecule when an oligoamine was used (Fig. 25). The electron transfer from the amine to the excited dye was proposed to be facile due to a bifunctional interaction between the dye and the adjacent free and protonated N-groups of the polyamine (*24*).

Photoredox polymers containing both dye and reducing groups were prepared by the reaction of poly(acrylic acid) (PAA) with dye and amine (en) in dry EtOH [92)] to get (*25*).

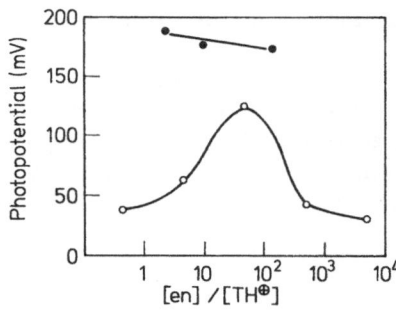

Fig. 26. Dependency of photopotential induced by photoredox polymer PAA/TH/en (●) and TH/en system (○) upon the mole ratio [en]/[TH]. The conc. of 25: 10^{-5} mol·1^{-1} based on repeating units of pendant TH^+; [TH^+]; 10^{-5} mol·1^{-1}, pH 8

The aqueous solution of the photoredox polymer induced photopotential without any other reagent. The photopotential induced by the polymer was much higher than that by the corresponding monomer system, and the value did not depend on the amine/dye ratio contrary to the monomeric system whose photopotential strongly depended on that ratio (Fig. 26). The presence of reductant near the dye and the isolated pendant dye structure were considered to be the cause for the high efficiency of the photoredox polymer.

A photoredox system of Fe^{2+} and polycation polymer pendant TH^+ (26) induced a much higher photopotential than the monomeric system [92]. A thin layer cell

composed of SnO_2 (nesa glass) and Pt electrodes was designed to construct a semi-dry photogalvanic cell [93]. A polymer gel such as PVA or poly(ethyleneimine) was used as a carrier for the redox couple, TH^+ and Fe^{2+}. The TH^+ polymer gel

from poly(epichlorohydrin), TH^+, and trimethylamine (27) had an output efficiency 60 times as high as the monomeric TH^+-Fe^{2+} system. A dry photochemical cell was constructed from two layers of cation and anion exchange resins adsorbing TH^+ and ascorbic acid, respectively [94].

5.4 New Photodiode Composed of a Polymer-metal Complex Film

Photodiodes utilize principally the photophysical process of semiconductors. The most typical juctions to attain photoinduced charge separation are shown in Fig. 27a–c. If a photoexcited compound (P) is arranged with donor and/or acceptor on an electrode as shown in Fig. 25 (d), it must work as a kind of *photodiode based on new principle of photochemical reaction*. A polymer film must be most promising to construct such photoconversion element.

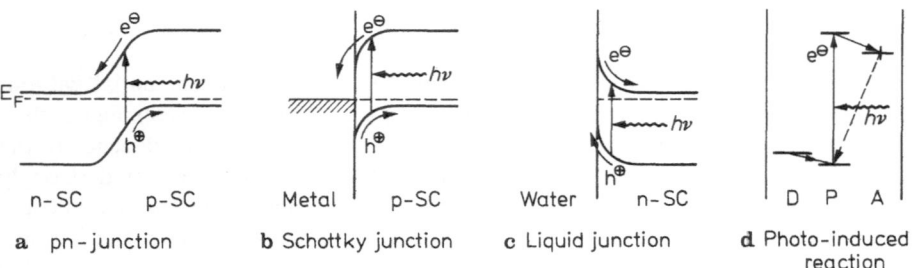

Fig. 27a—d. Photoinduced charge separation. SC: Semiconductor: D: Donor; A: Acceptor; P: Photochemical reaction center

The electron transfer from $Ru(bpy)_3^{2+*}$ to MV^{2+} is very rapid, the order being 10^8–10^9 M^{-1} s^{-1}. But the irradiation of an aqueous solution of both the compounds does not give a photoresponse at the electrode dipped in that solution, because the back electron transfer from $MV^{\ddagger}$ to $Ru(bpy)_3^{3+}$ is also very rapid; thereby no photochemical products are accumulated in the photostationary state (Scheme 6). However, when $Ru(bpy)_3^{2+}$ is modified on the electrode surface by utilizing polymer coating, the electrode shows photoresponse [95]. Such a modification of the electrode makes it photoresponsive, to afford a new type of photodiode.

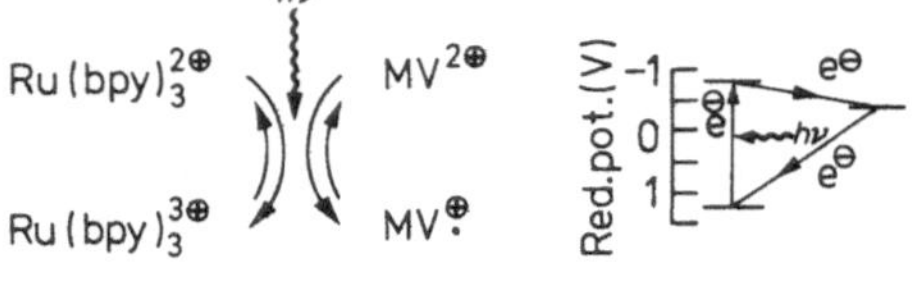

Scheme 6

The surface of a carbon electrode was at first coated with a thin film of an anionic polymer such as sodium poly(styrene-sulfonate) [95] or nafion [96] (thickness: thousand Å); then the cationic $Ru(bpy)_3^{2+}$ was adsorbed in the anionic layer electrostatically. The modification was also made by coating water insoluble polymer pendant $Ru(bpy)_3^{2+}$ (*8*) from its DMF solution [97]. These $Ru(bpy)_3^{2+}$/polymer modified electrode gave a photoresponse in the MV^{2+} solution with the Pt counter electrode [95-97]. The time-current behaviours induced by irradiation and cutoff of the light under argon are shown in Fig. 28. It is interesting to see that the direction of the photocurrent reversed at the electrode potential of ca. 0.4 V (vs. Ag—AgCl) under

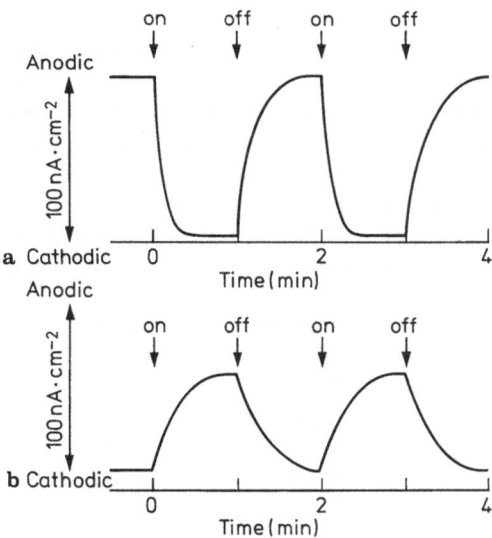

Fig. 28 a and b. Current changes induced by intermittent irradiation on Ru[P(St-Vbpy)]-(bpy)$_2^{2+}$ coated BPG (Basal plain pyrolitic graphite electrode) in 0.2 M CF$_3$COONa (pH 7) containing 10 mM MV^{2+} at (a) —0.4 V and (b) 0.8 V (vs. Ag—AgCl)

the condition of Fig. 28. MV^{2+} is adsorbed in the polymer layer and reacts there. Both the photochemical products, Ru(bpy)$_3^{3+}$ and MV^{+}, must react at the electrode. When the electrode potential is below 0.4 V, the reaction of Ru(bpy)$_3^{3+}$ to accept an electron from the electrode prevails (Fig. 29 a). When the electrode potential exceeds this value, however, the reaction of MV^{+} to donate electron to the electrode prevails (Fig. 29 b). Since MV^{+} is rapidly oxidized by oxygen during the photoelectro-

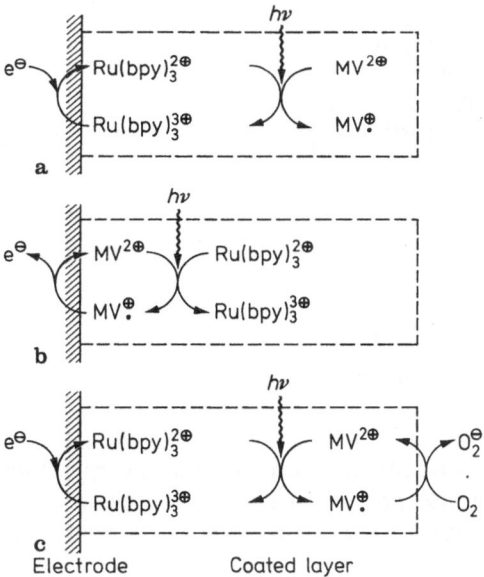

Fig. 29 a—c. Mechanism of photocurrent generation

chemical events, air reduces the concentration of MV^+, promotes the electrode reaction of $Ru(bpy)_3^{3+}$, and increases the cathodic photocurrent by about 100 times (Fig. 29c).

An anisotropic photoinduced electron flow is important to produce an efficient photocurrent. The anisotropic arrangement of the reaction components is necessary for that purpose. Polymer materials are very useful to achieve such arrangement. The arrangement of $Ru(bpy)_3^{2+}$ and MV^{2+} was attained by coating electrode at first with a polymer pendant Ru complex film and then with a polymer pendant MV^{2+} (28) film. The photocurrent induced by this bilayer coating system is shown in

Table 4 in comparison with that obtained from a mixture coated electrode [98]. The cathodic photocurrent at the electrode potential of -0.2 V generated by the bilayer coating is 4 times higher than that by the mixture coating. The anodic photocurrent at 0.7 V is, on the contrary, 13 times higher for the mixture coating than for the bilayer coating. The effect of the bilayer to produce an anisotropic photocurrent is calculated to be 52 times as high as the mixed system in Table 4.

Table 4. Bilayer modification composed of polymeric $Ru(bpy)_3^{2+}$ and MV^{2+}

	Bilayer		Mixed monolayer
	P[Ru(St-Vbpy)] $(bpy)_2^{2+}$	PMV^{2+}	P[Ru(St-Vbpy)] $(bpy)_2^{2+}$ $\overset{+}{PMV}{}^{2+}$
Appl. pot. (V vs. Ag-AgCl)	Photocurrent (nA/cm²)		
-0.2	238		59
0.2	72		13
0.7	-12		-159

$Ru[P(St-Vbpy)]$ $(bpy)_2^{2+}$; Ru 1 m M, 2 µl coated; Ru 11.8 nmol/cm²

Polymer coating thus not only makes electrodes photoresponsive by incorporating photoreactive function in that layer, but also provides a kind of new photodiode based on photochemical reaction by arranging the components in its multilayer structure (for example, see Fig. 30).

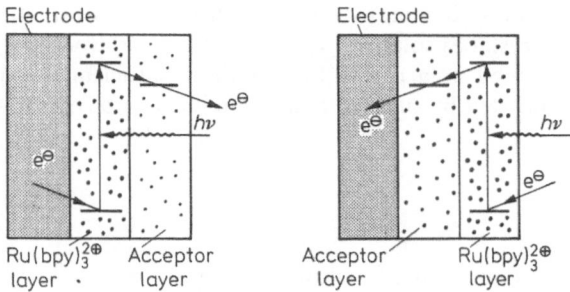

Fig. 30. New photodiode based on the photochemical reaction of the components incorporated in the polymer film

6 Miscellaneous

6.1 Solar Concentrator

Solar energy is difficult to be utilized due to its diluteness. In order to use it in a large scale, therefore, the irradiation must be collected. Especially when the conversion module is expensive (e.g., the Si solar cell), the collection and concentration of the solar energy is economically advantageous.

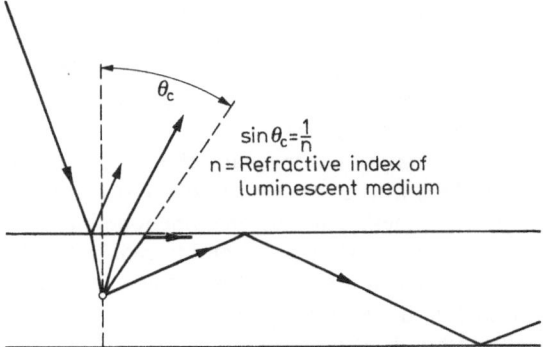

Fig. 31. Luminescent Solar Concentrator

A luminescent solar concentrator (LSC) is a planar layer of transparent polymer materials such as poly(methacrylic acid) containing luminescent dye molecules [99]. The dye absorbing light reemits light which can escape only at the layer edges (Fig. 31). The fraction of the light which is reemitted from the edges was calculated to be 70 to 80% when the refractive index of the polymer is 1.5 to 2. As for the fluorescent dyes,

29

the behaviours of xanthene compounds such as pyronin B (29) were studied. The efficiency of a conventional single-crystal Si solar cell was expected to be over 6.3% when coupled with LSC. Since a small area of the Si module is sufficient when LSC is used, the electricity cost generated by sunlight and Si module can be comparable with that by fossil fuels.

6.2 Energy Storage

Solar energy can be stored by isomerizing organic substance photochemically to an energy-rich unstable isomer. The photoisomerization of norbornadiene (NBD) to quadricyclane (QC) (Eq. (26)) is the most wellknown system to store solar energy one this principle [100a].

$$\text{(26)}$$

The QC is thermodynamically unstable relative to NBD, but a catalyst is needed to revert it to NBD. The reverse reaction of QC to NBD catalytically releases heat. The storage capacity of this is 21 kcal/mol. Since unsubstituted NBD has no absorption in visible region, a sensitizer such as aromatic ketones is required for the NBD isomerization by visible light. The efficiency and stability of various storage systems based on the photoisomerization were studied [100b].

The immobilization of the sensitizer and catalyst is especially effective, because contamination of the materials (NBD and QC) with a sensitizer or catalyst markedly lower the efficiency of this system. 4-(N,N-dimethylamino)benzophenone was immobilized on poly(styrene) (30) and silica gel to use it as insoluble sensitizer [101]. The polymer pendant sensitizer (30) was much more active than the monomeric compound when used in acetonitrile. Usually, the sensitizing activity of the sensitizer remained almost unchanged through immobilization, but sometimes decreased depending on their structure. As a catalyst of back reaction to release heat, Co(II)-tetraphenylporphyrine was anchored on poly(styrene) beads (31), and showed good activity in its immobilized form [102]. Activity decrease was observed after several times recyclings of the catalyst.

A flow type model utilizing a polymer pendant sensitizer and catalyst (Fig. 32) was constructed and operated.

For a large scale utilization of electricity generated by solar cells, storage of the electricity is most important, because solar irradiation is intermittent and unstable. The well known secondary battery using lead dioxide is not suited for a large scale storage. More efficient secondary battery of light weight, low cost, large capacity, and easy molding is desired to be developed for a large scale storage of electricity. Organic polymer materials are most suited for this purpose. Much attention was paid on *polymer batteries* utilizing poly(acetylene) electrodes [103].

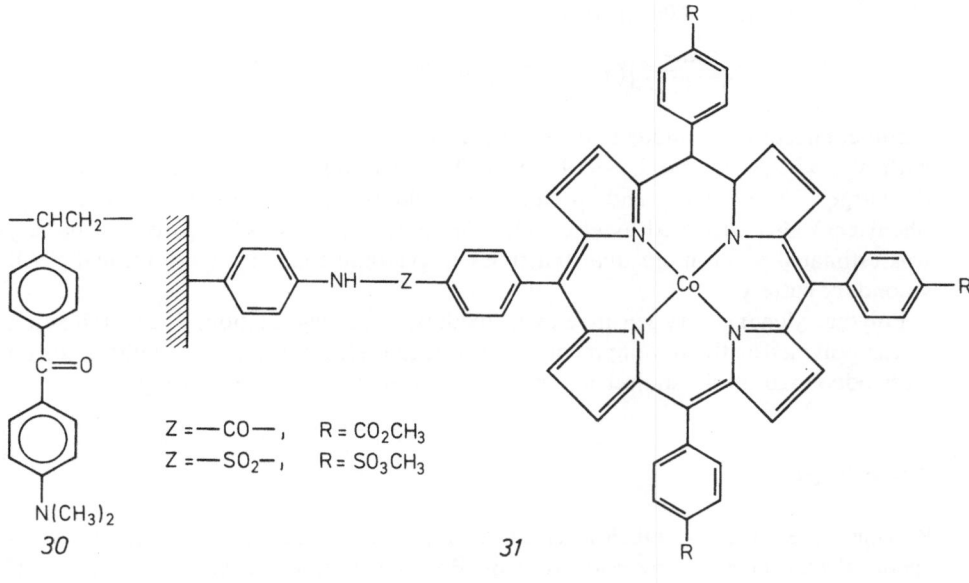

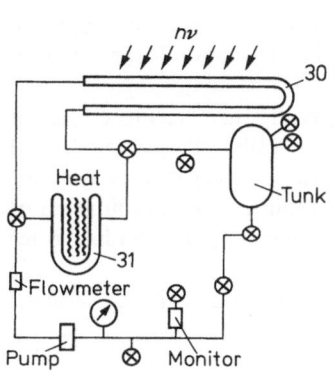

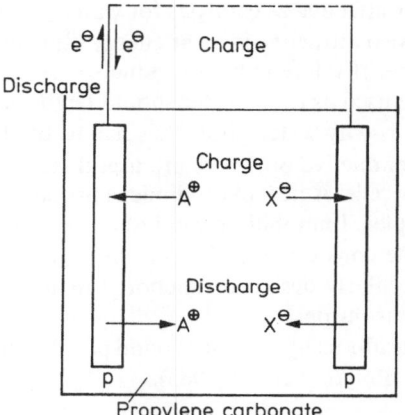

Fig. 32. Flow type solar energy storage system utilizing immobilized sensitizer (*30*) and catalyst (*31*)

Fig. 33. Polymer battery. AX; (n-Bu$_4$N)ClO$_4$, LiClO$_4$ etc.

A typical battery is composed of poly(acetylene) electrodes which are dipped in propylene carbonate containing tetraalkyl ammonium salt as doping material (Fig. 33). By charging poly(acetylene) is doped as represented by Eqs. (27) and (28).

$$(CH)_n + 0.06\,n\,(ClO_4^-)$$

$$\xrightarrow[\text{Discharge}]{\text{Charge}} [(CH^{+0.06})(ClO_4^-)_{0.06}]_n + 0.06\,ne^- \qquad (27)$$

$$(CH)_n + 0.06\,n(n\text{-}Bu_4N^+) + 0.06\,ne^-$$

$$\underset{\text{Discharge}}{\overset{\text{Charge}}{\rightleftharpoons}} [(n\text{-}Bu_4N^+)_{0.06}(CH^{-0.06})]_n \tag{28}$$

Upon connecting the anode and cathode in their doped state, a current is generated with $V_{oc} = 2.9$ V, and $I_{sc} = 1.9$ mA. The electrodes become undoped by this discharge. The charging and discharging could be repeated many times. Poly(p-phenylene) also was used as electrode. The power density (kW/kg) of the battery was estimated to be more than ten times larger than that of a conventional PbO_2 secondary battery.

Poly(acetylene) is very sensitive to air oxidation and degradation. The stabilization of the polymer or the development of a stable and efficient polymer material for an electrode is the future subject for the introduction of a polymer battery.

7 Outlook

Polymers are attracting much attention as functional materials to construct photo-chemical solar energy conversion systems. Polymers and molecular assemblies are of great value for a conversion system to realize the necessary one-directional electron flow. Colloids of polymer supported metal and polynuclear metal complex are especially effective as catalysts for water photolysis. Fixation and reduction of N_2 or CO_2 are also attractive in solar energy utilization, although they were not described in this article. If the reduction products such as alcohols, hydrocarbons, and ammonia are to be used as fuels, water should be the electron source for the economical reduction. This is why water photolysis has to be studied first.

Conductive polymers are useful semiconductors or coating materials to construct solar cells. A new photodiode is proposed to be made from a film of a polymer metal complex. Immobilized catalysts on polymers are used for solar energy storage systems.

The construction of solar energy conversion systems requires the combination of molecularly designed functional materials. Functional polymers play deciding roles for this purpose.

Results using polymer bound porphyrins as sensitizers are mentioned in: D. Wöhrle, Adv. Polym. Sci. *50* (1983).

8 Addendum

D. Wöhrle, guest editor

Additional to the results of the chapters 3.2 and 3.3, sensitizers bound at crosslinked polymers show high activity combined with high stability (G. Greber, W. Nußbaumer (Institut für Chemische Technologie Organischer Stoffe, Technische Universität Wien, A-1060 Wien Austria), Monatshefte für Chemie, in press):

Polymer bound tris-(2,2′-bipyridyl)-ruthenium-(II)-derivatives were synthesized by reaction of 4-(3-chloroformyl-propyl)-4′-methyl-2,2′-bipyridyl with crosslinked amino-groups containing poly(styrene) gel or crosslinked sucrose-methacrylate gels, and by subsequent reaction with cis-dichloro(2,2'-bipyridyl)ruthenium. These complexes were also used as photosensitizers for electron transfer from EDTA to methylviologen in aquous solution, and the reduced form of the viologen was used to reduce water to hydrogen analogous to Fig. 14. The hydrogen formation rates were lower than that of the comparable monomeric complexes, but 4 times higher than the best polymeric bound complexes known, and linear over a periode of more than 10 days. The hydrophilic or hydrophobic surrounding of the polymeric bound complexes did only influence the induction periode, but not their activities.

9 References

1 a. Grätzel, M.: Ber. Bunsenges. Phys. Chem. *84*, 981 (1980)
1 b. Lehn, J.-M.: Commentarii, *3*, 1 (1982)
2 a. Summary of 4th Internat. Conf. Photochem. Conversion and Storage of Solar Energy, 1982
2 b. Photochemical Conversion and Storage of Solar Energy, Connolly, J. S., Ed., Academic Press (1981)
3. Grätzel, M.: Acc. Chem. Res. *14*, 376 (1981)
4. Calvin, M.: Acc. Chem. Res.: *11*, 369 (1978)
5. Fendler, J. H.: J. Photochem. *17*, 303 (1981)
6. Calvin, M., Willner, I., Laane, C., Otvos, J. W.: J. Photochem. *17*, 195 (1981)
7. Foreman, T. K., Sobol, W. M., Whitten, D. G., J. Am. Chem. Soc. *103*, 5333 (1981)
8. Kuhn, H.: Pure Appl. Chem. *51*, 341 (1979)
9. Kaneko, M., Ochiai, M., Kinosita, K., Jr., Yamada, A.: J. Polym. Sci., Polym. Chem. Ed. *20*, 1011 (1982)
10. Kaneko, M., Yamada, A., Tsuchida, E., Kurimura, Y.: J. Polym. Sci., Polym. Lett. *20*, 593 (1982)
11. Kaneko, M., Yamada, A.: Kobunshi, *28*, 85 (1979)
12. Nishijima, T., Nagamura, T., Matsuo, T.: J. Polym. Sci., Polym. Lett. *19*, 65 (1981)
13. Govindjee, R.: Bioenergetics of Photosynthesis, Academic Press, 1975
14. Kaneko, M., Yamada, A.: Symp. Unsolved Probl. Polymer Chem. (Soc. Polym. Sci. Jpn.). pg. 21, 1976
15. Creutz, C., Sutin, N.: Proc. Nat. Acad. Sci. U.S. *72*, 2858 (1975)
16. Tazuke, S., Inoue, T., Tanabe, T., Hirota, S., Saito, S.: J. Polym. Sci., Polym. Lett. *19*, 11 (1981)
17. Tazuke, S., Tomono, H., Kitamura, N., Sato, K., Hayashi, N.: Chem. Lett. *1979*, 85
18. Takuma, K., Kajiwara, M., Matsuo, T.: Chem. Lett. *1977*, 1199
19 a. Turro, N. J., Grätzel, M., Braun, A. M.: Angew. Chem. *92*, 712 (1980)
19 b. Wallace, S. C., Grätzel, M., Thomas, J. K.: Chem. Phys. Lett. *23*, 359 (1973)
20. Tsutsui, Y., Takuma, K., Nishijima, T., Matsuo, T.: Chem. Lett. *1979*, 617
21. Brugger, P.-A., Infelta, P. P., Braun, A. M., Grätzel, M.: J. Am. Chem. Soc. *103*, 320 (1981)
22. Ford, W. E., Otvos, J. W., Calvin, M.: Nature, *274*, 507 (1978)
23. Matsuo, T., Itoh, K., Takuma, K., Hashimoto, K., Nagamura, T.: Chem. Lett., *1980*, 1009
24. Sudo, Y., Kawashima, T., Matsuo, T.: Chem. Lett., *1980*, 355
25. Fendler, J. H.: J. Photochem. *17*, 303 (1981)
26 a. Tundo, P., Kurihara, K., Kippenberger, D. J., Politi, M., Fendler, J. H.: Angew. Chem. *94*, 73 (1982)
26 b. Tundo, P., Kippenberger, D. J., Politi, M. J., Klahn, P., Fendler, J. H.: J. Am. Chem. Soc. *104*, 5352 (1982)
27. Atik, S. S., Thomas, J. K.: J. Am. Chem. Soc. *103*, 4279 (1981); *104*, 5868 (1982)
28. Clear, J. M., Kelly, J. M., Pepper, D. C., Vos, J. G.: Inorg. Chim. Acta, *33*, L139 (1979)
29. Shimidzu, T., Izaki, K., Akai, Y., Iyoda, T.: Polym. J. *13*, 889 (1981)
30. Kaneko, M., Nemoto, S., Yamada, A., Kurimura, Y.: Inorg. Chim. Acta, *44*, L289 (1980)
31. Kaneko, M., Yamada, A., Kurimura, Y.: Inorg. Chim. Acta, *45*, L73 (1980)
32. Ghosh, P. K., Spiro, T. G.: J. Am. Chem. Soc. *84*, 1319 (1980)

33.   Kaneko, M., Yamada, A., Nakajima, N., Tsuchida, E., Kurimura, Y.: Polym. Prepr. Jpn. *31*, 530 (1982)

34 a. Kurimura, Y., Shinozaki, N., Ito, F., Uratani, Y., Shigehara, K., Tsuchida, E., Kaneko, M., Yamada, A.: Bull. Chem. Soc. Jpn. *55*, 380 (1982)

34 b. Kaneko, M., Ochiai, M., Yamada, A., Kurimura, Y., Endo, W., Tsuchida, E.: Polym. Prepr. Jpn. *30*, 411 (1981)

35.   Matsuo, T., Sakamoto, T., Takuma, K., Sakurai, K., Ohsako, T.: J. Phys. Chem. *85*, 1277 (1981)

36.   Furue, M., Sumi, K., Nozakura, S.: J. Polym. Sci., Polym. Lett. *20*, 291 (1982)

37.   Sumi, K., Furue, M., Nozakura, S.: Polym. Prepr. Jpn. *31*, 2033 (1982)

38.   Sassoon, R. E., Rabani, J.: J. Phys. Chem. *84*, 1319 (1980)

39.   Sassoon, R. E., Rabani, J.: Israel J. Chem. *22*, 138 (1982); 4th Internat. Conf. Photochem. Conversion and Storage of Solar Energy, p. 230, 1982

40.   Kaneko, M., Yamada, A., Tsuchida, E., Kurimura, Y.: Polym. Prepr. Jpn. *31*, 1665 (1982)

41.   Kaneko, M., Awaya, N., Yamada, A.: Chem. Lett. *1982*, 619

42 a. Nishijima, T., Nagamura, T., Matsuo, T.: J. Polym. Sci., Polym. Lett. Ed. *19*, 65 (1981)

42 b. Ageishi, K., Endo, T., Okawara, M.: J. Polym. Sci., Polym. Chem. Ed. *19*, 1085 (1981)

43.   Lee, P. C., Matheson, M. S., Meisel, D.: Israel J. Chem. *22*, 133 (1982)

44.   Furue, M., Yamana, S., Phat, L., Nozakura, S.: J. Polym. Sci., Polym. Chem. Ed. *19*, 2635 (1981)

45.   Kaneko, M., Motoyoshi, J., Yamada, A.: Nature, *285*, 468 (1980).

46.   Kaneko, M., Yamada, A.: Makromol. Chem. *182*, 1111 (1981)

47.   Kaneko, M., Imamura, Y., Hamanishi, K., Yamada, A.: Kobunshi Ronbunshu, *39*, 665 (1982)

48.   Kaneko, M., Yamada, A.: Photochem. Photobiol. *33*, 793 (1981)

49.   Kurimura, Y., Nagashima, M., Takato, K., Tsuchida, E., Kaneko, M., Yamada, A.: J. Phys. Chem. *86*, 2432 (1982)

50.   Tsuchida, E., Nishide, H., Shimidzu, N., Yamada, A., Kaneko, M., Kurimura, Y.: Makromol. Chem. Rapid Commun. *2*, 621 (1981)

51.   Kawai, W.: Kobunshi Ronbunshu, *37*, 303 (1980)

52.   Kiwi, J., Grätzel, M.: Nature, *281*, 657 (1979)

53.   Toshima, N., Kuriyama, M., Yamada, Y., Hirai, H.: Chem. Lett. *1981*, 793

54.   Thiele, H., Levern, H. S. von: J. Colloid Sci. *20*, 679 (1965)

55.   Toshima, N., Yamada, Y., Hirai, H.: Polym. Prepr. Jpn. *30*, 416, 1500 (1981)

56.   Keller, P., Moradpour, A., Amouyal, E., Kagan, H.: J. Mol. Catal. *7*, 539 (1980)

57.   Akashi, M., Motomura, T., Miyauchi, N., O'Driscoll, K. F., Rempel, G. L.: Polym. Prep. Jpn. *30*, 1496 (1981)

58.   Lehn, J.-M., Sauvage, J.-P., Ziessel, R.: Nouv. J. Chim. *4*, 355 (1980)

59.   Henglein, A.: Angew. Chem., Int. Ed. *18*, 418 (1979)

60.   Henglein, A., Lilie, J.: J. Am. Chem. soc. *103*, 1059 (1981)

61.   Kalyanasundaram, K., Grätzel, M.: Angew. Chem. *91*, 759 (1979)

62.   Kaneko, M., Hamanishi, K., Yamada, A.: Ann. Meet. Chem. Soc. Jpn. p. 1128 (1981)

63.   Borgarello, E., Kiwi, J., Pelizzetti, E., Visca, M., Grätzel, M.: J. Am. Chem. Soc. *103*, 6324 (1981)

64.   Kaneko, M., Takabayashi, N., Yamada, A.: Chem. Lett. *1982*, 1647

65.   Buser, H. J., Schwarzenbach, D., Petter, W., Ludi, A.: Inorg. Chem. *16*, 2704 (1977)

66.   Fujishima, A., Honda, K.: Nature, *238*, 37 (1972)

67 a. Noufi, R., Frank, A. J., Nozik, A. J.: J. Am. Chem. Soc.: *103*, 1849 (1981)

67 b. Frank, A. J., Honda, K.: J. Phys. Chem. *86*, 1933 (1982)

68.   Ito, T., Shirakawa, H., Ikeda, S.: J. Polym. Sci., Polym. Chem. Ed. *12*, 11 (1974)

69.   Shirakawa, H., Louis, E. J., MacDiarmid, A. G., Chiang, C. K., Heeger, A. J.: J. Chem. Soc., Chem. Commun. *1977*, 578

70.   Chiang, C. K., Druy, M. A., Gau, S. C., Heeger, A. J., Louis, E. J., MacDiarmid, A. G., Park, Y. K., Shirakawa, H.: J. Am. Chem. Soc. *100*, 1013 (1978)

71.   Wegner, G.: Angew. Chem. *93*, 352 (1981)

72a. Chiang, C. K., Gau, S. C., Fincher, C. R., Jr., Park, Y. W., Mac Diarmid, A. G., Heeger, A. J.: Appl. Phys. Lett. *33*, 18 (1978)

72b. Chiang, C. K., Fincher, C. R., Jr., Park, Y. W., Heeger, A. J., Shirakawa, H., Louis, E. J., Gau, S. C., MacDiarmid, A. G.: Phys. Rev. Lett. *17*, 1098 (1977)

73. Shirakawa, H., Ikeda, S.: Kobunshi, *28*, 369 (1979)

74. Ghosh, A. K., Morel, D. L., Feng, T., Shaw, R. F., Rowe, C.: J. Appl. Phys. *45*, 230 (1974)

75. Kampas, F. J., Gouterman, M.: J. Phys. Chem. *81*, 690 (1977)

76a. Loutfy, R. O., Sharp, J. H.: J. Chem. Phys. *71*, 1211 (1979)

76b. Loutfy, R. O., Sharp, J. H., Hsiao, C. K., Ho, R., J. Appl. Phys. *52*, 5218 (1981)

77. Minami, N., Sasaki, K.: Autumn Meet. Chem. Soc. Jpn. p. 1502 (1981)

78. Matsuda, H., Minami, N., Sasaki, K.: Polym. Prepr. Jpn. *31*, 382 (1982)

79. Moriizumi, T., Kudo, K.: Appl. Phys. Lett. *38*, 85 (1981)

80. Misoh, K., Tasaka, S., Miyata, S., Sasabe, H., Yamada, A., Tanno, T.: Polym. Prep. Jpn. *31*, 384 (1982)

81. Donckt, E. V., Noirhomme, B., Kanicki, J.: J. Appl. Polym. Sci. *27*, 1 (1982)

82. Skotheim, T.: Appl. Phys. Lett. *38*, 712 (1981)

83. Glass, A. M., Linde, D., Negran, T. J.: Appl. Phys. Lett. *25*, 233 (1974)

84. Nakayama, T., Wachi, Y., Sasabe, H., Miyata, S.: Polym. Prepr. Jpn. *31*, 386 (1982)

85a. Simon, R. A., Ricco, A. J., Wrighton, M. S.: J. Am. Chem. Soc. *104*, 2031 (1982)

85b. Scotheim, T., Petersson, L.-G., Inganäs, O., Lundström, I.: J. Electrochem. Soc. *129*, 1737 (1982)

86. Noufi, R., Tench, D., Warren, L. F.: J. Electrochem. Soc. *127*, 2310 (1980)

87. Rajeshwar, K., Kaneko, M., Yamada, A.: J. Electrochem. Soc. *130*, 38 (1983)

88. Rabinowitch, E.: J. Chem. Phys. *8*, 551, 560 (1940)

89. Kaneko, M., Yamada, A.: Rep. Inst. Phys. Chem. Res. *52*, 210 (1976)

90. Kaneko, M., Yamada, A.: J. Phys. Chem. *81*, 1213 (1977)

91. Kaneko, M., Sato, S., Yamada, A.: Makromol. Chem. *179*, 1277 (1978)

92. Shigehara, K., Sano, H., Tsuchida, E.: Makromol. Chem. *179*, 1531 (1978)

93. Shigehara, K., Nishimura, M., Tsuchida, E.: Bull. Chem. Soc. Jpn. *50*, 3397 (1977)

94. Yoshida, M., Oshida, I.: Oyobutsuri, *33*, 34 (1964)

95. Kaneko, M., Ochiai, M., Yamada, A.: Makromol. Chem. Rapid Commun. *3*, 299 (1982)

96. Oyama, N., Yamaguchi, S., Kaneko, M., Yamada, A.: J. Electroanal. Chem. *139*, 215 (1982)

97. Kaneko, M., Yamada, A., Oyama, N., Yamaguchi, S.: Makromol. Chem. Rapid Commun. *3*, 169 (1982)

98. Kaneko, M., Yamada, A., Oyama, N.: Polym. Prepr. Jpn. *31*, 2049 (1982); Electrochim. Acta (in press)

99. Berman, E., Wildes, P.: 4th Internat. Conf. Photochem. Conversion and Storage of Solar Energy. p. 43, 1982

100a. Schwendiman, D. P., Kutal, C.: J. Am. Chem. Soc. *99*, 5677 (1977)

100b. Scharf, von H.-D., Fleischhauer, J., Leismann, H., Ressler, I., Schleker, W., Weitz, R.: Angew. Chem., *91*, 696 (1979)

101. Hautala, R. R., Little, J.: Interfacial Photoprocess: Applications to Energy Conversion and Synthesis. (M. S. Wrighton ed.), Adv. Chem. Ser. 184, p. 1 (1980)

102. King, R. B., Sweet, E. M.: J. Org. Chem. *44*, 385 (1979)

103. Macinnes, D., Jr., Druy, M. A., Nigrey, P. J., Nairns, D. P., MacDiarmid, A. G., Heeger, A. J.: J. Chem. Soc., Chem. Commun. *1981*, 317

Received February 2, 1983
H. J. Cantow (editor), D. Wöhrle (guest editor)

Polymer-Supported Phase Transfer Catalysts: Reaction Mechanisms

Warren T. Ford
Department of Chemistry, Oklahoma State University,
Stillwater, Oklahoma 74078, U.S.A.

Masao Tomoi
Department of Applied Chemistry, Faculty of Engineering, Yokohama National
University, Yokohama 240, Japan

Polymer-supported phase transfer catalysts are insoluble polymers having covalently bound functional groups active as catalysts for reactions between anions and neutral organic substrates. The active functional groups may be quaternary ammonium or phosphonium ions, crown ethers, cryptands, grafted poly(ethylene glycols), or analogues of dipolar aprotic solvents. The polymer most often used is polystyrene, but other synthetic polymers, silica gel, and alumina have been used also. The reactions are carried out in triphase mixtures of organic liquid, solid or aqueous inorganic salt; and solid catalyst. The chemical reactions take place within the polymer matrix or on the polymer surface. With heterogeneous catalysts the reaction rates may be limited by mass transfer of reactants to the catalyst surface, intraparticle diffusion of reactants to the active sites, or intrinsic reactivity at the active sites. An analysis of the experimental parameters which affect catalytic activity and their mechanisms of action is presented.

1 Introduction

Nature uses enzymes as catalysts, yet the number of polymeric catalysts in use in preparative chemistry is small. Surely if man could prepare catalysts with the high specificity and activity of most enzymes, they would replace the less specific, less active synthetic catalysts in use today.

This review will focus on polymer-bound catalysts in triphase systems, a small subset of polymeric catalysts. First let us briefly describe the breadth of the field of synthetic polymeric catalysts. Attempts have been made to mimic the action of enzymes with aggregates of small molecules such as micelles [1], complex single molecules such as the cycloamyloses and cyclic peptides, and synthetic linear polymers such as polypeptides, poly(vinylpyridines), and poly(ethyleneimines) [2]. The reactions most often studied have been ester hydrolyses, which serve as models of the enzyme-catalyzed cleavages of acyl carbon-heteroatom bonds [3]. (Only a few of the most recent reviews are cited here. For a more thorough list of reviews see Ref. 4.) Although enzymes are usually soluble polymers, most of the reactions they mediate likely take place in heterogeneous environments which differ markedly from bulk liquid water, such as the surfaces of membranes which enclose the living cell or the organelles within the cell.

In contrast, most experimental synthetic chemistry is carried out with soluble acidic and basic catalysts. On a large scale most often heterogeneous catalysts are used. These are commonly acids or metals on solid supports such as alumina, silica gel or zeolites. Synthetic polymers also have been used as catalyst supports for many years in the form of ion exchange resins, which serve as heterogeneous analogs of the common strong and weak acids and bases commonly used in homogeneous catalysis. The principal advantage of heterogeneous catalysts is the ease of separation of the catalyst from the reaction products. In batch reactions the catalyst is removed by filtration. In flow systems the reactant flows in and the product flows out of a fixed or fluidized bed of the catalyst [5,6]. The most common ion exchange resin catalysts are the sulfonated cross-linked polystyrenes used as heterogeneous strong acids. The largest scale use at present is in the manufacture of methyl t-butyl ether, a widely used anti-knock additive to gasoline, by acid-catalyzed addition of methanol to isobutylene. The sulfonic acid resins can catalyze virtually any reaction that is catalyzed in solution by sulfuric acid or p-toluenesulfonic acid. For reviews of their uses see Refs. 7–10.

Anion exchange resins have been used less often as catalysts. Weakly basic tertiary amine resins have been used for aldol condensations, [11,12] Knoevenagel condensations [13], Perkin reactions [14], and redistributions of chlorosilanes [15]. Quaternary ammonium ion resins can be used as a source of any reactive anion that can be bound to the resin. A recent review[16] cites many examples from the open literature. More examples, mainly from the patent literature, include hydroxide-catalyzed aldol condensations [17], additions of acetylenes to ketones [18], malonic ester syntheses [19], conjugate additions to ethyl acrylate [20] and acrylonitrile [21,22], additions of phenols to ethylene oxide [23,24], addition of hydrosulfide ion to epoxides [25], and additions of acrylate anions to epoxides [26]. For more extensive references to catalytic applications of anion exchange resins see Refs. 27, 28. Anion exchange resins also have served as supports for metal catalysts. A complex transition metal anion is bound to the resin and reduced to the catalytically active metal under the reaction conditions for hydrogenation [29–32], hydroformylation [33,34], or alkoxycarbonylation [35].

After the development of polystyrene-based ion exchange resins, more lightly cross-linked polystyrenes were introduced as supports for peptide [36,37] and poly-nucleotide [38] synthesis. With only 1% or 2% divinylbenzene these polymers swell and contract more than the common ion exchange resins, which were designed to be packed into colums for water treatment. The more swollen gels with higher solvent content are more fragile, compressible, and difficult to filter than the commercial ion exchange resins, but on a small scale they have given excellent yields in peptide syntheses. The same lightly cross-linked polymers were used later as supports for general organic synthesis [39-44] and for immobilized "homogeneous" transition metal catalysts [9,45-49].

The 1% and 2% cross-linked polystyrenes also have been used as supports for a variety of catalysts for organic reactions. Quaternized poly(4-vinylpyridine) gels have been used as catalysts for aspirin hydrolysis [50], and quaternary ammonium ion substituted polystyrene gels have been used as catalysts for decarboxylation of 6-nitro-benzisoxazole-3-carboxylate anion in aqueous buffer solutions [51]. Among the many other polystyrene-based catalysts are intercalated Lewis acids, photosensitizers, and groups such as imidazole that are active in enzymatic ester hydrolyses [39,52]. Most of these reactions are carried out in two phase systems which consist of solution and insoluble catalyst. Reports of polymer-bound quaternary onium ion and crown ether catalysts in triphase mixtures first appeared independently from the labo-ratories of Regen [53], Brown [54], and Montanari and Tundo [55]. A reactive anion from an aqueous salt solution and an organic substrate are brought together at the surface of or within the insoluble polymer, where the chemical reaction occurs. An early survey of the field [56] and a review comprehensive through part of 1981 [39] are available. Our review will concentrate on the mechanistic aspects of triphase catalysis for the purpose of understanding the fundamental chemical and physical parameters which control catalytic activity. This understanding can be applied to the design of more active catalysts and of catalysts for new chemical reactions.

Polymer-bound quaternary onium ions, crown ethers, cryptands and poly(ethylene glycols) are often called phase transfer catalysts because the structures of the catalytic sites are the same as those used as soluble phase transfer catalysts in liquid/liquid aqueous/organic mixtures [57-60]. Since the mechanisms of catalysis by soluble phase transfer agents resemble those of the polymer-bound analogs, a brief summary is presented here. A mixture of aqueous sodium cyanide and 1-chlorooctane fails to react even with vigorous mixing at 110 °C because the reactants fail to contact one another. A lipophilic quaternary ammonium ion catalyst such as methyltrioctylammo-nium chloride has a higher affinity for the organic than for the aqueous phase. The onium ion attracts cyanide into the organic phase. Once cyanide ion and 1-chloro-octane are in the same phase, they react rapidly at 110 °C (Eq. (1)).

$$\text{n-C}_8\text{H}_{17}\text{Cl} + \text{NaCN(aq)} \rightarrow \text{n-C}_8\text{H}_{17}\text{CN} + \text{NaCl} \tag{1}$$

Similarly, a lipophilic crown ether is partitioned into the organic phase. It's complexing ability serves to transfer salts with alkali cations into the organic phase. The anions in the organic phase are poorly solvated and highly reactive. The overall reactivity in phase-transfer-catalyzed nucleophilic displacement reactions thus is a function of both the partition coefficient for extraction of the reactive anion into the organic phase and the intrinsic reactivity of the anion with the organic substrate. Under some conditions the rate of transfer of the reactive anion into the organic phase may be a rate-limiting factor, as shown by dependence of observed reaction rates on the speed of stirring the two phase mixture.

Scheme 1

$$C_6H_5CH_2CN + NaOH(aq) \rightleftharpoons C_6H_5\overset{\ominus}{C}HCN, \overset{\oplus}{Na} + H_2O$$

$$C_6H_5\overset{\ominus}{C}HCN, \overset{\oplus}{Na} + RBr \longrightarrow C_6H_5CHRCN + NaBr$$

Scheme 2

$$CHCl_3 + NaOH(aq) \rightleftharpoons \overset{\ominus}{C}Cl_3, \overset{\oplus}{Na} + H_2O$$

$$\overset{\ominus}{C}Cl_3, \overset{\oplus}{Na} \longrightarrow {:}CCl_2 + NaCl$$

Phase-transfer-catalyzed reactions which use hydroxide ion to generate a reactive intermediate, such as alkylation of phenylacetonitrile (Scheme 1) or dichlorocyclopropanation of olefins with chloroform (Scheme 2), are thought to proceed at the aqueous/organic interface rather than within the organic phase. The principal evidence for an interfacial reaction is the poor extractability of hydroxide ion into organic solvents with even the most lipophilic cations, and the high activity of catalysts such as benzyltriethylammonium chloride, which are not as lipophilic as the catalysts most active in nucleophilic displacement reactions. The upper limits of the hydroxide ion concentration in the organic phase and of the phenylacetonitrile concentration in the aqueous phase are too low to account for the observed reaction rates if the reactions take place entirely in one phase. Therefore the reaction most likely occurs at the aqueous/organic interface. Vigorous mixing is required for fast reactions.

Phase transfer catalysts bound to insoluble supports offer both advantages and disadvantages compared with the analogous soluble catalysts. The heterogeneous catalysts can be separated from reaction mixtures by filtration in batch processes or used in fixed or fluidized bed reactors. Many of the soluble quaternary ammonium and phosphonium salts used as phase transfer catalysts have modest surfactant properties and can cause emulsions which make isolation of the product from the organic phase difficult. Most lipophilic quaternary ammonium ions also have bacteriostatic activity which could decrease the effectiveness of sewage treatment. The activities of the heterogeneous catalysts usually are lower because of diffusional limitations on the reaction rates. Also the preparation of the heterogeneous catalyst is an added expense that can be recovered only by extensive recycling of the catalyst. (With various approximations we estimate that the cost of manufacturing a polystyrene-based benzyltrimethylammonium ion catalyst, the most common type of anion exchage resin, is five to ten times higher than that of monomeric benzyltrimethylammonium chloride.

Thus five to ten catalytic cycles would be required to recover the cost of the catalyst if that' was the only variable in the cost of an organic chemical manufacturing process).

Most heterogeneous catalysts are bound to inorganic supports such as alumina and silica gel, which are chemically stable at higher temperatures and are more physically durable than most polymers [5, 6]. When feasible chemical manufacturing processes are run at high temperatures where the reaction rates are fast enough that only the surface catalytic sites are active. In contrast, polymer-supported catalysts are generally employed at lower temperatures, and they utilize active sites both on the surface and within the solvent-swollen polymer matrix. Metals and strong acids are the catalysts most often used on inorganic supports. The active sites in polymer-supported catalysts are usually organic or organometallic functional groups with higher specificity and lower thermal stability. The variety of polymers and methods of functionalization available make it possible to control the swelling properties, the hydrophilic/lipophilic balance, and the pore structure of the support in the design of catalysts for specific reaction processes. Inorganic supports have a lower number of catalytic sites per unit weight than the more highly substituted polymers. Those sites are in a highly polar environment on inorganic supports unless the surface is modified chemically by lipophilic groups. This review will discuss catalysts supported on silica gel and alumina only when they have been used in phase transfer processes analogous to the uses of polymer-supported catalysts.

2 General Mechanism of Catalysis

Three fundamental processes can limit overall reaction rates in heterogeneous catalysis: mass transfer of the reactants from the bulk liquid phase(s) to the surface of the solid catalyst, diffusion of the reactants from the catalyst surface to the active site, and the intrinsic reaction at the active site [61, 62]. Each of these processes depends on one or more experimental parameters, as shown in Fig. 1.

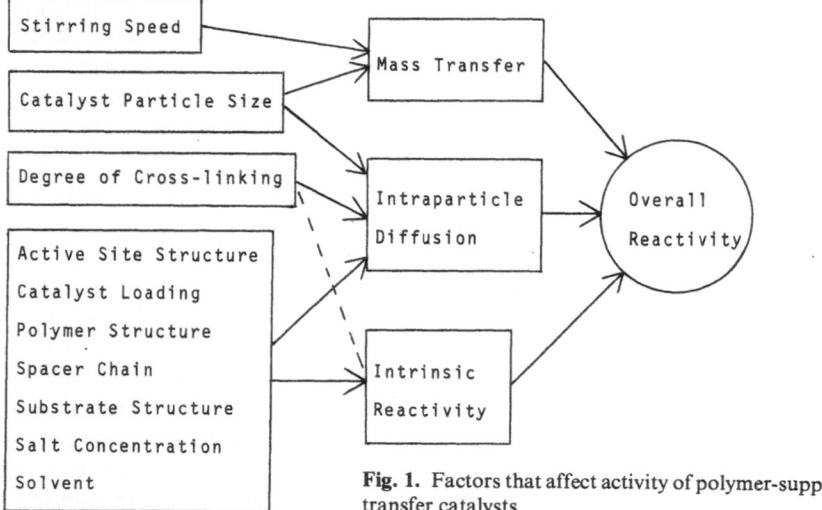

Fig. 1. Factors that affect activity of polymer-supported phase transfer catalysts

2.1 Mass Transfer

Mass transfer can take place by diffusion through the bulk liquid, and it is faciliated by agitation of the reaction mixture. No matter how well the reaction mixture is agitated, however, there is still a quiet layer of liquid on the surface of the solid catalyst, known as the Nernst diffusion layer to electrochemists. Agitation of the reaction mixture reduces the thickness of the quiet film on the catalyst surface. With no agitation the mass transfer of reactants to the catalyst surface would occur by diffusion through the liquid. Vigorous mixing can reduce the thickness of the film diffusion layer to a fraction of the dimensions of the catalyst particle [61, 62]. Fig. 2 shows how mixing efficiency affects the thickness of the film diffusion layer. The general description of mass transfer applies both to stirred batch reactors and to flow reactors. Since the best characterized polymer-supported phase transfer catalysts have been used in the form of spherical beads in stirred batch reactions, only that special case of mass transfer will be considered further.

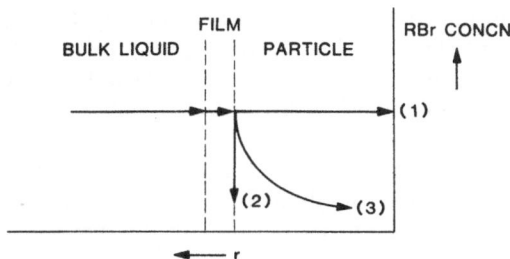

a FAST MIXING

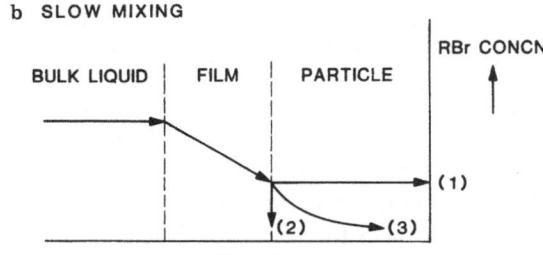

b SLOW MIXING

Fig. 2. Reactant concentration as a function of distance from the center of a catalyst particle for fast mixing (A) and slow mixing (B). In both figures, (1) represents a reaction rate limited by intrinsic reactivity at the active site, (2) represents a reaction rate limited by mass transfer, and (3) represents a reaction rate limited by a combination of intraparticle diffusion and intrinsic reactivity. (Reprinted with permission from Ref. [73]. Copyright 1981 American Chemical Society)

2.2 Intraparticle Diffusion

After the reactants reach the catalyst particle surface, they must be transported to an active site before reaction can occur. All polymeric catalysts under consideration here swell to some degree in liquids. The number of active sites on the particle surface is small, so reactants must diffuse through the solvent-swollen matrix to find active sites. The effect of intraparticle diffusion on catalytic activity can range from almost nil to so high that only surface catalytic sites are active. Intraparticle diffusion is never rate-limiting by itself; it always acts together with intrinsic reactivity at the active sites. The effect of intraparticle diffusional limitation of the reaction rates can be described by Eq. (2), in which k is the intrinsic rate constant, M_c is the molar equivalents of catalyst, λ is the partition coefficient for reactant A between the catalyst and the external liquid, C_{AS} is the concentration of A at the catalyst surface, and η is the effectiveness factor, the factor by which intraparticle diffusion reduces the observed rate from the intrinsic reaction rate [7].

$$-dN_A/dt = kM_c\lambda C_{AS}\eta \tag{2}$$

At one extreme diffusivity may be so low that chemical reaction takes place only at suface active sites. In that case η is equal to the fraction of active sites on the surface of the catalyst. Such a polymer-supported phase transfer catalyst would have extremely low activity. At the other extreme when diffusion is much faster than chemical reaction $\eta = 1$. In that case the observed reaction rate equals the intrinsic reaction rate. Between the extremes a combination of intraparticle diffusion rates and intrinsic rates controls the observed reaction rates as shown in Fig. 2, which profiles the reactant concentration as a function of distance from the center of a spherical catalyst particle located at the right axis. When both diffusion and intrinsic reactivity control overall reaction rates, there is a gradient of reactant concentration from C_{AS} at the surface, to a lower concentration at the center of the particle. The reactant is consumed as it diffuses into the particle. With diffusional limitations the active sites nearest the surface have the highest turnover numbers. The overall process of simultaneous diffusion and chemical reaction in a spherical particle has been described mathematically for the cases of ion exchange catalysis, [63-65] and catalysis by enzymes immobilized in gels [66,67]. Many experimental parameters influence the balance between intraparticle diffusional and intrinsic reactivity control of reaction rates with polymer-supported phase transfer catalysts, as shown in Fig. 1.

Macroporous [68-71] polymers present an additional diffusion step in catalysis. They have permanent pores created during synthesis in addition to the micropores, which are the spaces between chains in the polymer network filled by solvent in solvent-swollen form. Transport of a reactant from the surface of a macroporous catalyst particle to an active site may proceed first through the liquid-filled macropores to the internal surface of the catalyst particle and then through the polymer matrix. A mathematical treatment of catalysis kinetics of macroporous ion exchange resins is available [72].

2.3 Intrinsic Reactivity

The structure of the active site of polymer-supported phase transfer catalysts has been studied more than any other experimental parameter. In this section we describe those features most vital for catalytic activity without citing specific examples. Sections 3 and 4 provide the details.

A large cationic radius generally increases the nucleophilic reactivity of the anion in an ion pair. A small cationic radius results in stronger electrostatic attraction and lesser reactivity of the anion. Catalysts with the active site separated from the polymer backbone by an aliphatic chain have higher activity than those prepared by quaternization of a chloromethylpolystyrene with a tertiary amine or tertiary phosphine, leaving only one carbon atom between the active site and the aromatic ring.

The percent ring substitution ($\%$ RS) of the polymer with active sites affects catalytic activity. Polystyrenes with $<25\%$ RS with lipophilic quarternary onium ions are swollen in triphase mixtures almost entirely by the organic phase. Water reduces the activity of anions by hydrogen bonding. In most triphase nucleophilic displacement reactions onium ion catalysts with $<25\%$ RS are highly active, and those with $>40\%$ RS, such as most commercial ion exchange resins, are much less active. However, low $\%$ RS is not critical for the reactions of hydroxide ion with active methylene compounds, as commericial ion exchange resins work well in alkylation of active nitriles.

The affinity of the polymer-bound catalyst for water and for organic solvent also depends upon the structure of the polymer backbone. Polystyrene is nonpolar and attracts good organic solvents, but without ionic, polyether, or other polar sites, it is completely inactive for catalysis of nucleophilic reactions. The polar sites are necessary to attract reactive anions. If the polymer is hydrophilic, as a dextran, its surface must be made less polar by functionalization with lipophilic groups to permit catalytic activity for most nucleophilic displacement reactions. The $\%$ RS and the chemical nature of the polymer backbone affect the hydrophilic/lipophilic balance. The polymer must be able to attract both the reactive anion and the organic substrate into its matrix to catalyze reactions between the two mutually insoluble species. Most polymer-supported phase transfer catalysts are used under conditions where both intrinsic reactivity and intraparticle diffusion affect the observed rates of reaction. The structural variables in the catalyst which control the hydrophilic/lipophilic balance affect both activity and diffusion, and it is often not possible to distinguish clearly between these rate limiting phenomena by variation of active site structure, polymer backbone structure, or $\%$ RS.

3 Triphase Catalysis by Quaternary Ammonium and Phosphonium Ions

3.1 Nucleophilic Displacement Reactions

3.1.1 Mass Transfer

The method and speed of mixing affect observed rates in triphase catalysis when the chemical reactions are fast. For the reaction of 1-bromooctane in toluene with aqueous sodium cyanide (Eq. (3))

$$\text{n-C}_8\text{H}_{17}\text{Br} + \text{NaCN(aq)} \xrightarrow{1} \text{n-C}_8\text{H}_{17}\text{CN} + \text{NaBr(aq)} \tag{3}$$

typical rate dependences on stirring speed are shown in Fig. 3 [73]. The experiments were performed with an overhead mechanical stirrer and common Teflon blade in

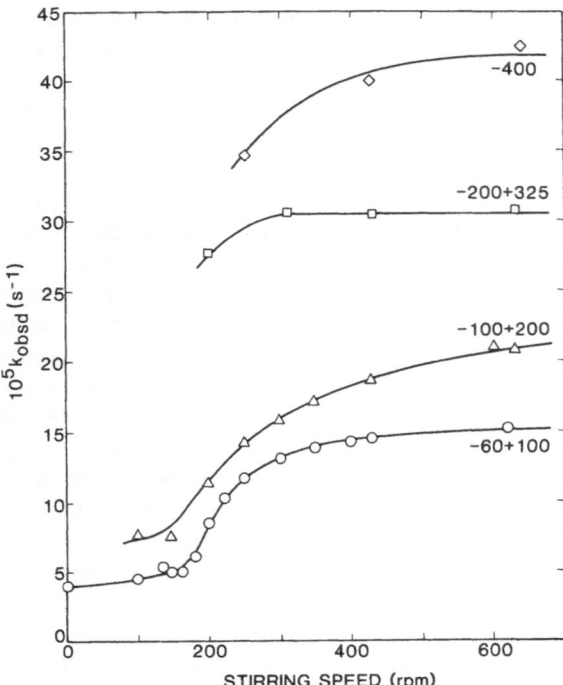

Fig. 3. Dependence of k_{obsd} on stirring speed and particle size (mesh) of 2% CL catalysts *1* for reaction of 1-bromooctane in toluene with 10 molar equiv of NaCN in water and 0.02 molar equiv of catalyst at 90 °C. (Reprinted with permission from Ref. [73]. Copyright 1981 American Chemical Society)

a 100 ml round-bottomed flask containing 40 g of aqueous phase, 24 g of organic phase, and 0.4 g of 2% CL polystyrene 17% RS with benzyltri-n-butylphosphonium ions (*1*).

$$P \!-\!\!\!\bigcirc\!\!\!-\!\!\!\bigcirc\!\!-CH_2\overset{\oplus}{P}(n\text{-}C_4H_9)_3$$

1

With no stirring the catalyst resides at the liquid/liquid interface and the reaction still proceeds, but at a slower rate. Fig. 3 appears almost the same as figures which show stirring speed effects on rates of liquid/liquid phase transfer catalysis [58]. Mass transfer is the only rate-limiting factor that depends on mixing. Thus the observed reaction rates must be mass transfer dependent at all stirring speeds below those that give the maximum rates in Fig. 3. To determine if still more vigorous agitation would increase the rates, otherwise identical kinetic experiments were carried out with vibromixing and with ultrasonic mixing. Much finer dispersions of the organic phase in the aqueous phase were achieved, but no further increases in rate were observed. All subsequent kinetic experiments on reactions of cyanide ion with 1-bromooctane were carried out with 600 rpm stirring, shown to give the maximum rates [73].

It is critical to remove mass transfer as an experimental variable if one wishes to compare activities of polymer-supported catalysts as a function of any other vairable.

Ideally this should be done by performing the experiments with efficient mixing to get out of the mass transfer limited domain, but if the mixing method and speed are carefully standardized, it is possible to learn qualitatively about other experimental variables even when mass transfer is partly rate-limiting. If mass transfer effects are not carefully standardized, variations in mixing from one experiment to the next could give highly misleading interpretations of results.

Rates of the reaction of 1-bromooctane with aqueous potassium iodide (Eq. (4))

$$n\text{-}C_8H_{17}Br + KI \xrightarrow{1} n\text{-}C_8H_{17}I + KBr(aq) \tag{4}$$

catalyzed by 1 depend on the speed of magnetic stirring [74]. Using a 12 mm stirring bar in a 4 ml flask, 1.5 g of aqueous phase, 1.3 g of organic phase and 15 mg of catalyst the rate constants reached a maximum at about 1000 rpm. Use of a flask with vertical creases and a magnet 20 mm long, comparible in size with the flask bottom, gave more turbulent mixing and increased rates. Most of the rates were determined under the less turbulent conditions. Subsequently it was found that a break-water sheet in the flask gave the same maximum rate at 600 rpm that was attained at ca. 1000 rpm without the break-water sheet [75].

Another approach to control of stirring speed as a variable was taken by Regen [76] in a study of the kinetics of the reaction of n-decyl methanesulfonate in toluene with saturated aqueous sodium chloride catalyzed by 1 (Eq. (5)).

$$n\text{-}C_{10}H_{21}O_3SCH_3 + NaCl(aq) \xrightarrow{1} n\text{-}C_{10}H_{21}Cl + NaO_3SCH_3 \tag{5}$$

Unstirred mixtures were studied to avoid complications due to reproducibility of mixing methods. The unstirred rates were stated to be identical to those obtained with > 750 rpm magnetic stirring. The absence of stirring dependence is due to the much slower experimental rates. The Tomoi and Ford [73] kinetics of cyanide displacement on 1-bromooctane and the Montanari [74] kinetics of iodide displacement on 1-bromo-octane were carried out with pseudo-first-order half lives of 0.5 to 3 h, whereas the Regen [76] kinetics of chloride displacement on n-decyl methanesulfonate had half lives of 12 h or more. When chemical reaction is that slow, mass transfer can occur by diffusion, and observed reaction rates do not depend on mass transfer. Thus the study of very slow reactions is another way to avoid the problems of mass transfer limited kinetics and to study other variables in catalysis. Subsequent papers by Regen [77,78] have reported use of vigorous 1500 rpm stirring in reactions performed on the same slow time scale.

3.1.2 Intraparticle Diffusion

Intraparticle diffusion limits rates in triphase catalysis whenever the reaction is fast enough to prevent attainment of an equilibrium distribution of reactant throughout the gel catalyst. Numerous experimental parameters affect intraparticle diffusion. If mass transfer is not rate-limiting, particle size effects on observed rates can be attributed entirely to intraparticle diffusion. Polymer % cross-linking (% CL), % ring substitution (% RS), swelling solvent, and the size of reactant molecule all can affect both intrinsic reactivity and intraparticle diffusion. Typical particle size effects on the

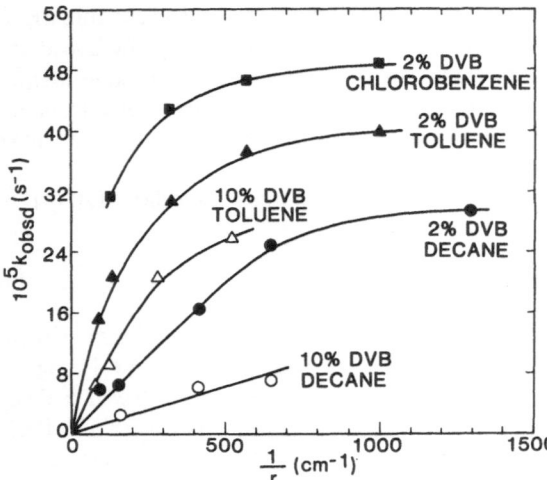

Fig. 4. Dependence of k_{obsd} on mean catalyst particle size and solvent. Stirring speed is 600–650 rpm, and other experimental conditions are the same as in Fig. 3. (Reprinted with permission from Ref. [73]. Copyright 1981 American Chemical Society)

reaction of 1-bromooctane with aqueous sodium cyanide catalyzed by *1* are shown in Fig. 4 [73]. All experiments were carried out with 600 rpm mechanical stirring to remove mass transfer as a rate limiting factor. A linear dependence of observed rate constants on inverse particle radius (1/r), indicates that the rate constants are directly proportional to the catalyst surface area, and that only active sites on the surface are used because chemical reaction is much faster than intraparticle diffusion. Such conditions are likely for the reactions employing the larger particles and decane as the solvent shown in Fig. 4. As particle sizes decrease to a mean diameter of 20—30 μm, the slopes of the 1/r plots in Fig. 4 appear to approach zero. When reaction rates do not depend on particle size, they are controlled only by intrinsic reactivity. Fig. 4 demonstrates that the rates of the prototypical triphase nucleophilic displacement of Eq. (3) carried out with half-lives between 0.25 and 3 h, depend on a combination of intraparticle diffusion and intrinsic reactivity.

With the less active polystyrene-bound benzyltrimethylammonium ion catalysts *2* (17% RS),

$$\text{P}\!\!-\!\!\bigcirc\!\!-\!\!\bigcirc\!\!-\!\!CH_2\overset{\oplus}{N}(CH_3)_3$$

2

the rate constants for reaction of 1-bromooctane with cyanide ion were smaller by factors of at least ten [73]. The much slower reactions did not depend on catalyst particle size or stirring speed, demonstrating that one can avoid intraparticle diffusional as well as mass transfer limitations by carrying out the kinetics over a long time.

Polystyrene-supported quaternary phosphonium ions with spacer chains between the active site and the aromatic ring (*3, 4*, 17–19% RS)

$$P-\bigcirc-(CH_2)_n\overset{\oplus}{P}(n-C_4H_9)_3$$

$$3: n = 4$$
$$4: n = 7$$

$$P-\bigcirc-CH_2NH\overset{O}{\overset{\|}{C}}(CH_2)_{10}\overset{\oplus}{P}(n-C_4H_9)_3$$

$$5$$

were more active than the benzylphosphonium ions *1* studied first in the reaction of 1-bromooctane in toluene with aqueous sodium cyanide (Eq. (13)) [79, 80]. As expected with faster reactions, the rate constants strongly depended on particle size. Similar dependence of rate constants on the particle size of phosphonium ion catalyst *5* (36% RS, 17% CL) was observed for the reaction of iodide ion with 1-bromooctane (Eq. (4)) [75], but no such particle size dependence was observed with 1% CL. Two particle sizes differing by a factor of ten were tested for catalysis of the reaction between chloride ion and n-decyl methanesulfonate (Eq. (5)) with a low activity 52% RS catalyst *1* [76]. Even though the half-lives of the reactions were 0.5 to 4 days, the smaller catalyst particles were more active by a factor of two.

No other papers have considered carefully the effects of catalyst particle size on activity. Comparisons of catalysts with different particle sizes could be misleading. Fortunately, most investigators have used a single batch of chloromethylated polystyrene to prepare their catalysts, and the subsequent comparisons of activities with different active site structures are likely valid.

Variation in % CL of the catalyst support most likely affects intraparticle diffusion more than it affects intrinsic reactivity. Increased cross-linking causes decreased swelling of the polymer by good solvents. Thus the overall contents of the gel become more polystyrene-like and less solvent-like as the % CL is increased. Fig. 5 shows the

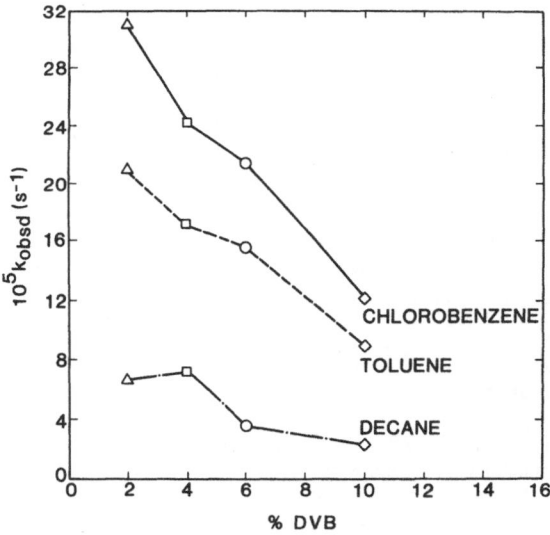

Fig. 5. Effect of solvent and % CL on k_{obsd}. Stirring speed is 600–650 rpm, and other experimental conditions are the same as in Fig. 3. (Reprinted with permission from Ref. [73]. Copyright 1981 American Chemical Society)

effects of polymer cross-linking and solvent on rate constants of reaction between 1-bromooctane and aqueous sodium cyanide catalyzed by *1* [73]. The rate constants decreased with increased % CL from 2% to 10% divinylbenzene using both good and poor solvents. The swelling ratios (swollen volume/dry volume) for the 2% CL catalyst at 25 °C were chlorobenzene 3.0, toluene 2.2, and decane 1.0. The observed reaction rates in Fig. 5 also clearly decreased with decreased swelling of the catalyst as a function of solvent, but the effects of the solvents on the intrinsic reactivity are unknown. It is not possible at this time to attribute the data in Fig. 5 entirely to intra-particle diffusional effects.

Poly(4-vinylpyridine) resins *6* cross-linked with 9.6% divinylbenzene and 68–82% alkylated also have been tested as catalysts for reaction of 1-bromooctane with cyanide ion [81]. The catalytic activities depended on the organic solvent in the order benzene > toluene > o-dichlorobenzene. No swelling data were reported, so it is not known if the activities correlate with the swollen volumes of the catalysts.

$$\text{P}-\text{C}_6\text{H}_4-\overset{\oplus}{\text{N}}-n\text{-}C_nH_{2n+1} \qquad 6:\ n = 2, 4, 8, 12, 16$$

Comparison of polystyrene-supported phosphonium ion catalysts *1* in the reaction of 1-bromooctane with iodide ion showed a 4.5% CL catalyst to be only half as active as a 2% CL catalyst [74]. Use of decane, toluene and o-dichlorobenzene as solvents gave rate constants that increased as the swelling ability of the solvent increased. Swelling ratios were measured at 90 °C, the reaction temperature.

Polystyrene catalysts *1*, 1% to 20% CL with divinylbenzene, have been tested in the reaction of chloride ion with n-decyl methanesulfonate (Eq. (5)) [77]. The catalyst

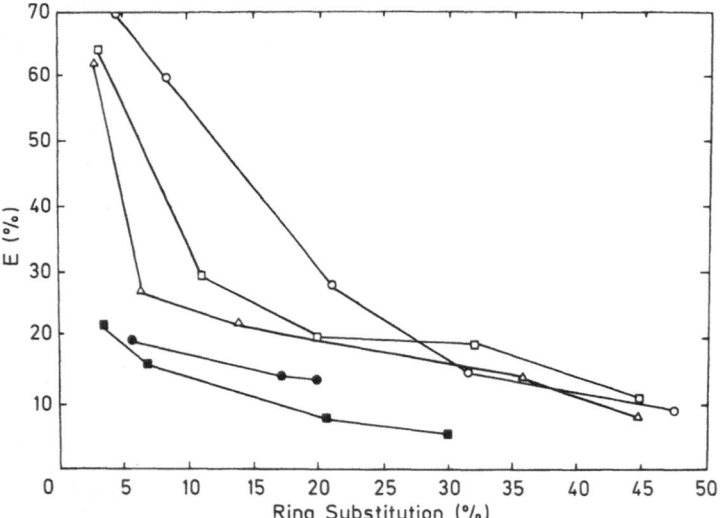

Fig. 6. Plot of catalyst efficiency, E, as a function of % RS for 1% (○), 2% (□), 5% (△), 10% (●), and 20% (■) CL catalysts *1* with 1500 rpm magnetic stirring. $E = k_{obsd}(1)/k_{obsd}$ ($C_6H_5CH_2PBu_3^+Cl^-$) for equimolar amounts of catalysts. (Reprinted with permission from Ref. [77]. Copyright 1981 American Chemical Society)

efficiency, expressed as the rate constant with the insoluble catalyst divided by the rate constant with the analogous soluble phase transfer catalyst benzyltri-n-butyl-phosphonium chloride, decreased as % CL increased as shown in Fig. 6. No experimental rate constants were published, but presumably the conditions involved half-lives of 12 h or more as in previous papers [76].

Observed rate constants depend upon the size of the reactant molecule which must diffuse through the gel to an active site. The relative reactivities of 1-bromoalkanes (C_4 to C_8) with aqueous sodium phenoxide have been studied with a variety of polystyrene-bound catalysts [82]. With 1 % CL, 17 % RS phosphonium ion catalysts 1 the range of rate constants was only $C_4Br/C_8Br = 1.4$. The analogous ammonium ion catalyst 2 gave $C_4Br/C_8Br = 2.07$. Half-lives of the faster reactions for the two catalysts were 0.25 and 4 h respectively. It is surprising that the less active catalyst showed the greater size selectivity, since intraparticle diffusion should be more important with the more active catalyst. Regen and Nigam [82] considered also possible selective adsorption of alkyl bromides as a possible cause of selectivity, but that possibility was ruled out by adsorption experiments. A likely explanation is that the tri-n-butylphosphonium ion catalyst 1 swells much more than the trimethylammonium ion catalyst 2 in toluene, as shown by swelling and ^{13}C NMR line width measurements [83], and consequently intraparticle diffusion is more of a limitation in 2 in spite of its lower activity.

Substrate selectivities in reactions of aqueous sodium cyanide with alkyl halides in toluene and 17 % RS onium ion catalysts are shown in Table 1 [84]. The data are particularly instructive about how intraparticle diffusion affects reactions that occur

Table 1. Rates of Reaction of Alkyl Halides in Toluene with Aqueous NaCN at 90 °C. (Reprinted with permission from Ref. [84]. Copyright 1981 American Chemical Society)

Catalyst		$10^5 k_{obsd}$, s^{-1}			
% DVB	Size, mesh	n-C_8-$H_{17}Br^a$	n-C_{16}-$H_{33}Br^a$	C_6H_5-CH_2Br^b	C_6H_5-CH_2Cl^a
Benzyltri-n-butylphosphonium Catalysts^c					
2	$-100 + 200$	21		125	
2	$-200 + 325$	31	13		
2	$-325 + 400$	38	16	400	
2	-400	40	17	580	
10	$-100 + 200$	9		47	
10	$-325 + 400$	26		170	
Benzyltrimethylammonium Catalysts^d					
2	$-60 + 100$	1.4	0.21	44	5.9
2	$-325 + 400$	1.4	0.52	112	6.1
10	$-60 + 100$	1.0		26	6.3
10	$-325 + 400$	1.8		99	7.8

^{a} 2.0 mol % catalyst;
^{b} 0.5, 1.0, or 2.0 mol % catalyst. Rate constants are normalized to those for 2.0 % catalyst, assuming k_{obsd} is directly proportional to catalyst concentration;
^{c} 600–650-rpm mechanical stirring;
^{d} 420–440-rpm mechanical stirring

over a range of time scales. With the phosphonium ion catalyst *1*, 2 % and 10 % CL, 1-bromooctane reacted faster than 1-bromohexadecane by a factor of 2.4 with catalysts of three different particle sizes, and rate constants increased as particle sizes decreased. With the less active, less swollen, ammonium ion catalysts *2* the C_8/C_{16} selectivity was even greater, but the 2 % CL catalyst gave no dependence of rate constant on particle size. This suggests that intraparticle diffusion of 1-bromoctane was fast relative to intrinsic reactivity in catalyst *2* and was no longer a rate limiting parameter. The more slowly diffusing 1-bromohexadecane, however, still showed a rate constant dependence on particle size of *2*. Comparison of cross-linking and particle size effects on the rates of reaction of benzyl bromide and benzyl chloride with cyanide ion (Table 1), shows that the more reactive benzyl bromide was quite sensitive to the parameters which affect intraparticle diffusion, but the less reactive benzyl chloride was almost completely insensitive to them. Apparently the benzyl chloride reactions were controlled mainly by intrinsic reactivity.

The most important point about the alkyl halide reactivities in triphase catalysis is that the reactions which have the highest intrinsic rates are the most likely to be limited by intraparticle diffusion. The cyanide ion reactions which showed the greatest particle size and cross-linking dependence with 1-bromooctane had half-lives of 0.5 to 2 h and with benzyl bromide had half-lives of 0.13 to 0.75 h. The reactions of 1-bromooctane and of benzyl chloride which were insensitive to particle size and cross-linking had half-lives of 14 h and 3 h respectively. Practical triphase liquid/liquid/solid catalysis with polystyrene-bound onium ions has intraparticle diffusional limitations.

It is not known whether the intraparticle diffusional limitations involve ion diffusion or organic reactant diffusion. Most kinetic experiments have been performed with large excesses of the ionic reagent in concentrated aqueous solutions under pseudo first order conditions. Such a method simplifies the heterogeneous kinetics, but it hides kinetic dependences on the reactive anion. Our initial experiments [73,85] with 1-bromooctane and aqueous sodium cyanide (Eq. (3)) were carried out with a tenfold excess of cyanide under the assumption that the rate of ion exchange would be much faster than the rate of exchange of product 1-cyanooctane for reactant 1-bromooctane. Moreover, stop-action photographs of the stirred reaction mixtures showed that the catalyst particles and the organic droplets were dispersed in a continuous aqueous phase [73]. Apparently the catalyst was in constant contact with the aqueous phase and only intermittent contact with the organic phase, which suggested that mass transfer of the organic reactant might have been slower than mass transfer of the cyanide ion. Subsequently we discovered that catalysts recovered from reaction mixtures after 57—86 % consumption of the 1-bromooctane contained bromide ion at 62 to 77 % of the ion exchange sites [86]. The sites farthest from the catalyst surface are likely the slowest to exchange ions and have the lowest turnover numbers, so the presence of 62–77 % bromide ion does not necessarily mean significant rate limitation by slow ion exchange. In fact, reasonable pseudo-first-order kinetics were usually observed to 57–86 % conversion. If there were significant reductions in the numbers of the most active sites, there should have been decreases in the instantaneous first order rate constants with time, which were not observed [73].

Catalysts *1* (17 % RS) also were recovered from both stirred and unstirred reaction mixtures containing aqueous sodium chloride and n-decyl methanesulfonate

(Eq. (5)) [76]. The stirred reaction had a half life of 0.8 h, and the resin was fully maintained in chloride form. The much slower unstirred reaction, however, showed a 33 % decrease in the chloride content of the resin, indicative of slow mass transfer of chloride ion. Intraparticle diffusion of chloride ion was not a limitation, but mass transfer was under unstirred conditions.

Another indication of slow ion exchange is the detection of a reaction product resulting from the original counter ion of the polymer-bound onium ion catalyst. When the reaction of 1-bromooctane with cyanide ion was studied with macroporous 50% and 75% CL, 3% RS phosphonium ion catalysts *1*, analyses of the reaction mixtures after low conversion detected 1-chlorooctane, even though the catalyst was conditioned in the aqueous phase for 60 min at 90 °C before starting the kinetic determinations [86]. Chloride ion and cyanide ion have approximately the same affinity for most polystyrene-based ion exchange resins, so the presence of chloride ion in the resin in spite of a 500-fold excess of cyanide in the external aqueous solution must have been due to slow ion exchange. Approximate calculations of the rates of ion exchange in 2% CL, 17% RS non-macroporous and 10% CL, 11% RS macroporous catalysts were made from observed reaction rates and ion contents of recovered catalysts [86]. The estimated intraparticle diffusivities of cyanide and bromide ions were smaller under triphase conditions than under conventional ion exchange conditions in aqueous solutions by factors of 5 to 50. Similar conclusions of slow ion exchange were drawn about the reaction of 1-iodooctane with aqueous sodium chloride and alkylated poly(4-vinylpyridinium bromide) catalysts *6* [87]. In the early stages of the reaction 1-bromooctane was detected as a product. However, the catalyst was not conditioned in the aqueous phase before adding the 1-iodooctane.

Recent evidence for rate limitation by intraparticle diffusion of ions was found in the effect of % RS on activity of polystyrene-bound phosphonium ions *1*, *3*, and *4* [80]. The rate constants for cyanide displacement on 1-bromooctane increased as % RS increased from 7–9% to 17–19% and decreased again with >30% RS (Fig. 7).

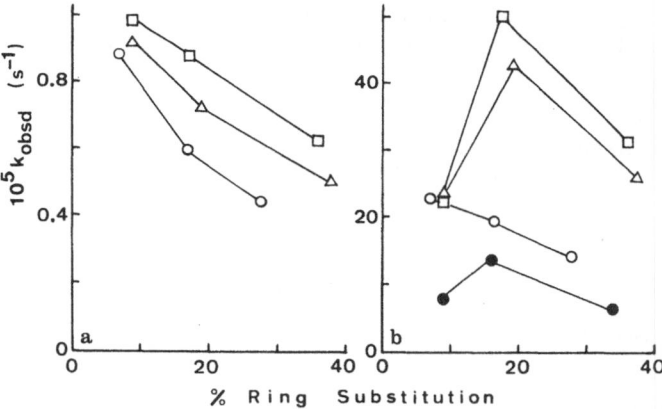

Fig. 7. Dependence of k_{obsd} on substrate structure and % RS for reactions of 1-chlorooctane (left) and 1-bromooctane (right) in toluene with aqueous NaCN at 90 °C and 0.02 molar equiv of 100/200 mesh catalysts 2% CL with divinylbenzene; catalyst *1* (○), catalyst *3* (△), catalyst *4* (□), 10% CL catalyst *1* (●). (Reprinted with permission from Ref. [80]. Copyright 1982 John Wiley and Sons. Inc.)

The former increase in activity may be attributed to faster ion exchange. The latter decrease may be attributed to either slower transport of organic reactant or to decreased activity of a more higly hydrated catalyst. These experiments will be discussed in more detail in Sect. 3.1.3.

Both reaction rates as a function of % RS and rates of ion exchange have been determined for the reaction of aqueous sodium acetate with 1-bromooctane and with benzyl chloride [88]. At 90 °C using 2% CL pnosphonium ion catalyst 4 reaction rates were higher with 17% RS than with 9% RS or 36% RS. Exchange of acetate ion for bromide ion in the catalysts under triphase conditions (aq NaOAc, toluene, 30 °C, 650–700 rpm stirring, 74–149 μm particles) reached limiting conversion after 5 min with the 17% RS and 36% RS catalysts but was much slower with the 9% RS catalyst. These are the only results available that demonstrate a dependence of ion exchange rate on % RS, but the phenomenon is likely to be general.

3.1.3 Intrinsic Reactivity

Parameters which affect the intrinsic activity of polymer-supported phase transfer catalysts are listed in Fig. 1.

The structure of quaternary onium ion active sites was one of the first parameters to be studied [53,89]. Valid comparisons of active site structure can be made only if all other variables are kept constant. Ideally mass transfer and intraparticle diffusion should not be rate limiting factors. With 7–10% RS, 2% CL polystyrenes the tri-n-butyl and tri-n-octylammonium ions 7 and 8

$$7 : R = n\text{-}C_4H_9$$
$$8 : R = n\text{-}C_8H_{17}$$

were equally active and the tri-n-butylphosphonium ion 1 was more active in stirred reactions of 1-bromooctane with aqueous potassium iodide [55]. At the end of a long spacer chain both tri-n-butylammonium and tri-n-butylphosphonium ions showed high activity (Table 2). The much higher activity of lipophilic quaternary onium ions

Table 2. Effect of Onium Ion Structure on Activity for Reaction of 1-Bromooctane in Toluene with Iodide Ion at 90 °C [74]

Catalyst, 2% CL	mequiv/g	$k \times 10^6$ s^{-1}
	1.2	20.4
	1.1	330
	1.4	272

than of the more hydrophilic trimethylammonium ions shown in Table 2 is well known in liquid/liquid phase transfer catalysis with soluble onium salts [57-60]. The early data of Regen [89] indicated that a wide variety of 8–10% RS quaternary ammonium ions on 2% CL polystyrene had nearly the same activity for cyanide displacement on 1-bromooctane (Eq. (13)) in unstirred mixtures. Half-lives estimated by us from data in the original paper [89] were 0.5 to 2.5 h at 110 °C. The half-lives suggest in light of more recent evidence that such small effects of active site structure were observed because the reactions were strongly limited by mass transfer and intraparticle diffusion. In magnetically stirred mixtures of 1-bromopentane and aqueous potassium cyanide at 110 °C with catalysts bound to 2% CL polystyrene the phosphonium ion 9 (32% RS) gave a rate constant twice that of the ammonium ion 10 (29% RS), but phosphonium and ammonium ion catalysts 11 (16% RS) and 12 (26% RS) had nearly equal activity [90,91].

$$\boxed{P} - \bigcirc - (CH_2)_n \overset{\oplus}{X} (n\text{-} C_4H_9)_3$$

9 : X = P, n = 2
10 : X = N, n = 2
11 : X = P, n = 3
12 : X = N, n = 3

Overall, there is no large difference in activity between lipophilic phosphonium and ammonium ions, although the phosphonium ions appear to be more active in locations close to the polymer backbone.

Table 3. Effect of % Ring Substitution on Rate of Reaction of 1-Bromooctane in Toluene with Cyanide Ion at 90 °C [89]

Catalyst, 2% CL	% RS	k_{rel}
13	10	0.7
13	21	0.7
2	10	1.0
2	46	0.005
2	76	<0.005

The % ring substitution of the polymer is a critical factor in catalytic activity. Its importance was demonstrated clearly in Regen's first full paper on triphase catalysis [89]. Catalysts 2 and 13 (2% CL) were active for cyanide displacement on 1-bromooctane (Eq. (3)) only at 21% or lower RS (Table 3). Commerical anion exchange resins, polystyrenes highly substituted as benzyltrimethylammonium ions 2 or benzyldimethyl-(2-hydroxyethyl)ammonium ions 14, were inactive.

$$\boxed{P} - \bigcirc - CH_2 \overset{\oplus}{N} (CH_3)_2 R$$

13: R = n-C₄H₉
14: R = CH₂CH₂OH

Similar results have been found for reaction of 1-bromopentane with aqueous potassium cyanide [91]. Regen's active catalysts had swelling ratios of 2.3 in toluene and 1.7–1.8 in water, demonstrating their ability to imbibe both organic and ionic reactants, whereas the inactive ion exchange resins had swelling ratios of only 1.3 in toluene [89]. The ion exchange polymer must be nonpolar to absorb more organic phase than aqueous phase. A detailed study of the effect of % RS on reaction of n-decyl methanesulfonate with chloride ion (Eq. (5)) catalyzed by 1 (Fig. 6) shows the same trend [77]. However, catalysts with too low % RS may be less active because of slow ion exchange [80,88].

Not all nucleophilic displacement reactions require lightly substituted onium ion catalysts for activity. For alkylation of 2-naphthoxide ion with benzyl bromide (Eq. (6)) 40–100% RS, 2% CL polystyrene catalysts 15 and 16 work well [54]. A 51% RS catalyst 11 gave good yields in reactions of anionic oxygen and sulfur nucleophiles with alkyl halides [91].

$$15 : n = 11$$
$$16 : n = 5$$

The concentration of salt in the aqueous phase can have a large effect on the activity of polymer-bound onium ion phase transfer catalysts. Ohtani and Regen [92] studied the reaction of n-decyl methanesulfonate with chloride ion (Eq. (5)) under triphase catalytic, triphase stoichiometric, and biphase stoichiometric conditions using 52% and 17% RS, 1% CL phosphonium ion catalysts 1. The biphase experiments employed 0.45 mmol of catalyst in chloride form with a toluene solution of 0.02 mmol of the n-decyl methanesulfonate in the absence of an external aqueous phase. With both 52% and 17% RS catalysts incremental addition of water to the catalyst, raising the hydration state from 0 to 18.5 water molecules per active site, reduced the pseudo first order rate constants by a factor of more than ten. An external aqueous phase containing no sodium chloride gave hydration states of 297 water molecules per ionic site in the 52% RS catalyst and 96 in the 17% RS catalyst (triphase stoichiometric conditions) and reduced the rate constants to 0.002 and 0.08 times those observed with concentrated aqueous NaCl (the triphase catalytic conditions). The triphase catalytic rate constants were 0.15 and 0.23 times the biphase stoichiometric rate constants with no hydration, and were slightly higher than the biphase rate constants obtained at hydration levels equal to those under the triphase catalytic conditions. The similar rate constants under triphase catalytic and comparable biphase stoichiometric conditions rule out any large intraparticle diffusional limitation to the triphase reactions (which had half-lives of 4.4 and 1.6 h). The overall results demonstrate that hydration of the active site reduces activity and suggest that the reason for lower activity of the 52% RS catalyst is its greater hydration under triphase conditions.

Substantial variations of the organic solvent used in triphase catalysis with poly-styrene-bound onium ions have been reported only for the reactions of 1-bromo-octane with iodide ion (Eq. (4)) [74] and with cyanide ion (Eq. (3)) [73]. In both cases observed rate constants increased with increasing solvent polarity from decane to toluene to o-dichlorobenzene or chlorobenzene. Since the swelling of the catalysts increased in the same order, and the experiments were performed under conditions of partial intraparticle diffusional control, it is not possible to determine how the solvents affected intrinsic reactivity.

Table 4. Effect of Spacer Chains on Activity for Reaction of 1-Bromooctane in Toluene with Iodide Ion at 90 °C [93]

Catalyst, 2% CL	mequiv/g[a]	$t_{1/2}$ min
P—⟨○⟩—$CH_2[NHCO(CH_2)_{10}]_n \overset{\oplus}{P}(n\text{-}C_4H_9)_3$		
1 : $n = 0$	2.0	180
5 : $n = 1$	1.1	72
17 : $n = 2$	1.4	54
18 : $n = 3$	ca. 1.1	48

[a] From Ref [74]; assumed the same as in Ref. [93]

Polystyrenes with the onium ion sites separated from the aromatic ring with spacer chains are more active than those with only a single carbon between charged nitrogen or phosphorus and the ring [93]. Table 4 shows the effects of long spacer chains on the triphase reaction of 1-bromooctane with iodide ion (Eq. (4)). There is only a small gain in activity by the addition of two or three 11-atom chains instead of just one 11-atom chain (*5*). Replacement of the most common one carbon link (*1*) with a three carbon chain between phosphorus and the ring as in 2% CL catalyst *11* produced a threefold increase in activity for triphase reaction of cyanide ion with 1-bromopentane under conditions where both intraparticle diffusion and intrinsic activity probably were rate limiting [91].

Spacer chain catalysts *3*, *4*, and *19* have been investigated under carefully controlled conditions in which mass transfer is unimportant (Table 5) [80]. Activity increased as chain length increased. Fig. 7 shows that catalysts *3* and *4* were more active with 17–19% RS than with 7–9% RS for cyanide reaction with 1-bromooctane (Eq. (3)) but not for the slower cyanide reaction with 1-chlorooctane (Eq. (1)). The unusual behavior in the 1-bromooctane reactions must have been due to intraparticle diffu-sional effects, not to intrinsic reactivity effects. The aliphatic spacer chains made the catalyst more lipophilic, and caused ion transport to become a limiting factor in the case of the 7–9% RS catalysts. At >30% RS organic reactant transport was a rate limiting factor in the 1-bromooctane reactions [80]. In contrast, the rate constants for the 1-chlorooctane reactions were so small that they were likely limited only by intrinsic reactivity. (The rate constants were even smaller than those for the analogous reac-tions of 1-bromooctane and of benzyl chloride catalyzed by polystyrene-bound benzyl-

Table 5. Effect of Short Spacer Chains on Activity in Reaction of 1-Bromooctane in Toluene with Cyanide Ion at 90 °C [80]

Catalyst, 2% CL 100/200 mesh	mequiv/g	% RS	$k \times 10^5$ s^{-1}
$\boxed{P}$—◯—$(CH_2)_n \overset{\oplus}{P}(n\text{-}C_4H_9)_3$ *1* : $n = 1$	1.08	17	20
3 : $n = 4$	1.09	19	43
4 : $n = 7$	0.99	17	50
$\boxed{P}$—◯—$CH_2O(CH_2)_3 \overset{\oplus}{P}(n\text{-}C_4H_9)_3$ *19*	1.01	17	31

trimethylammonium ions)[84]. Hence the intrinsic activity of spacer chain catalysts *1*, *3*, and *4* for a prototypical nucleophilic displacement increased as the length of the the spacer chain increased, and the effects of spacer chain length on intraparticle diffusion are ones of hydrophilic/lipophilic balance. Fig. 8 shows the affinities of toluene, methanol, and water for catalysts *1*, *3*, and *4*. The differences between the solvents were greatest with > 30% RS. Water and methanol uptake decreased as the length of the spacer chain increased. Figs. 7 and 8 indicate that a minimum water uptake is required to prevent ion exchange from becoming a rate limiting factor in the reaction of cyanide ion with 1-bromooctane.

Polymer supports different from polystyrene show different response to % RS in triphase catalysis of nucleophilic displacements. Poly(4-vinylpyridines) 66–71% alkylated with C_8, C_{12}, and C_{16} chains (*6*) give high yields of cyanide displacement on 1-bromooctane and on 1-iodooctane and of bromide/iodide exchange of the 1-halooctanes [81,87]. An 81.6% RS n-butyl-quaternized poly(4-vinylpyridine) (*6*)

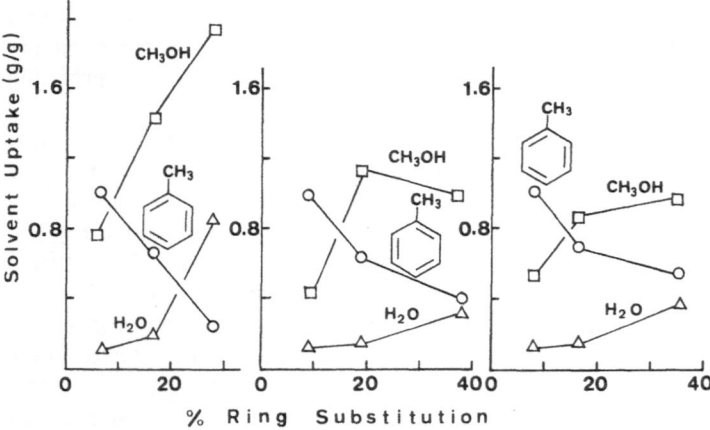

Fig. 8. Effects of % RS and spacer-chain length on the amount of solvents imbibed into 100/200 mesh 2% CL catalysts at 25 °C on the basis of g solvent/g dry catalyst. Left, catalyst *1*; center, catalyst *3*; right, catalyst *4*. (Reprinted with permission from Ref. [80]. Copyright 1982 John Wiley and Sons, Inc.)

gave two times the rate constant found with a 15.5% isobutyl-substituted analog in iodide displacement on 1-bromooctane [87].

Poly(styrene-g-ethyleneimine) in the form of macroporous 3% CL beads with linear polyethyleneimine blocks was 42% quaternized with 1-bromobutane (20) [94].

$$\text{P}-\bigcirc-\text{CH}_2\!\!\left(\overset{\oplus}{\text{N}}\text{CH}_2\text{CH}_2\right)_{\!n}$$
$$\underset{20}{} (n\text{-}C_4H_9)_2$$

The resulting catalyst was highly active for cyanide and acetate ion displacements on 1-bromobutane. As expected, soluble low molecular weight quaternary ammonium salts and a soluble quaternized linear poly(ethyleneimine) were even more active, presumably because they had no mass transfer and intraparticle diffusional limitations. These catalysts had a much higher density of charged sites (at least within the micro domains of the poly(ethyleneimine)) than any of the other active quaternary ammonium ion catalysts reported for nucleophilic displacement reactions.

Dextran anion exchangers modified with lipophilic substituents were active catalysts for cyanide and iodide displacements on 1-bromoalkanes [95]. The catalyst structures 21–23

$$\text{dextran}-\text{OSiCH}_2\text{CH}_2\text{CH}_2\overset{\oplus}{\text{N}}-\text{Bu} \quad \text{Cl}^{\ominus}$$

with substituents OMe/OMe on Si and Bu/Bu on N

$$\underset{21}{}$$

$$\text{dextran}-\text{OCH}_2\text{CH}_2\overset{\oplus}{\text{N}}-\text{CH}_2\text{CHCH}_3 \quad \text{Cl}^{\ominus}$$

with substituents Et/Et on N and OR

22: R = SiMe$_3$
23: R = CH$_2$Ph

had hydroxypropylated, trimethylsilylated, and benzylated dextrans as the polymer supports. All three were active triphase catalysts. In contrast, the quaternary ammonium ion resin QAE Sephadex and a resin prepared by the method used for 22, but not hydroxypropylated, were inactive. The dextrans not modified with lipophilic groups were too polar to act as catalysts, either because they prevented transport of the organic reactants to the active sites or because large amounts of water within the polymer matrix inhibited the reaction.

3.2 Alkylation of Active Methylene Compounds

Active methylene compounds have more than one activating group such as carbonyl, cyano, sulfonyl, or aryl bound to a methylene carbon. Bases such as hydroxide ion easily remove a proton to form a reactive carbanion. The most widely studied example is the alkylation of phenylacetonitrile (Scheme 1). The abstraction of the proton is generally the rate limiting step.

With soluble quaternary ammonium salts as catalysts the reaction is thought to take place at the aqueous/organic interface because a) the solubilities of quaternary ammonium hydroxides in organic solvents are too low to account for the observed reaction rates, and b) the most active catalysts are benzyltriethylammonium and similar ions which have lower affinities for organic solvents than do many more lipophilic quaternary ammonium salts [96-99]. Quaternary phosphonium ions are not used as catalysts because they decompose rapidly in strong bases such as 50% aqueous sodium hydroxide [100]. Commercial quaternary ammonium ion anion exchange resins [101], soluble quaternized poly(vinylpyridines) [102], and benzyltrimethyl-ammonium ions bound to polystyrenes of varied % RS and % CL [103] have been used as catalysts for alkylation of phenylacetonitrile (Eq. (7)).

$$C_6H_5CH_2CN + n\text{-}C_4H_9Br + NaOH \rightarrow$$
$$C_6H_5CH(n\text{-}C_4H_9)CN + NaBr + H_2O \tag{7}$$

Mass transfer and diffusional effects have been investigated only with polystyrene supports [103]. The apparatus and mixing methods employed were the same as those described earlier for reaction of 1-bromooctane with aqueous sodium cyanide [73]. With excess 1-bromobutane as the organic solvent, a limited amount of phenylacetonitrile, 50% aqueous sodium hydroxide, and 1–4 mole % of the catalysts 2, rate constants increased as the stirring speed was increased until a constant value was reached at ca. 500 rpm [103,104]. Vibromixing did not increase the reaction rate. Rate constants depended upon particle size of 2% CL catalysts for 15% to 50% RS catalysts. Pseudo first order half lives were 7 to 24 min at 80 °C. The fast reactions were limited by mass transfer unless fast mixing was used, and intraparticle diffusion was always a limiting factor.

The rate dependence on the method of mixing the triphase system provided a particularly dramatic example of the importance of conditioning the catalyst under swelling conditions [104]. When the 2% CL, 15% RS catalyst was stirred with the aqueous phase and 1-bromobutane for 60 min before addition of phenylacetonitrile, the rate was only one eighth as fast as when the same experiment was performed by a) stirring the catalyst with the aqueous phase for 50 min, b) adding the phenylacetonitrile, and c) after another 10 min adding the 1-bromobutane. The only difference between the two experimental methods was the order of addition of the organic reactants. The catalyst swelled to 3.0 times its dry volume in 5 min in phenylacetonitrile at 80 °C, whereas it required 12 h to reach a final volume of 1.4 times its dry volume in 1-bromobutane. Addition of 1-bromobutane to the phenylacetonitrile-swollen catalyst in the proportions used in kinetic experiments caused a contraction from 3.0 to 2.0 times dry volume over a period of 30 min, a time longer than the half life in kinetic experiments. Pseudo first order kinetics were observed within limits of experimental error, although the swelling experiments indicate that the observed reaction rate constant would have been smaller if equilibrium swelling had been reached before the kinetic data were obtained.

Only limited information is available on the effect of active site structure on activity of polymer-supported quaternary ammonium ion catalysts for alkylation of phenyl-acetonitrile (Eq. (7)). Dou and co-workers [101] reported approximately the same activity for six different type 1 (2) and type 2 (14) Dowex [105] ion exchange resins. Conversions of 50–70% were obtained in 10 h at 70 °C using 0.1 mol each of phenyl-acetonitrile and 1-bromobutane, 40 ml of 50% NaOH and 1–4 g (3.5–18 mequiv) of resin. The triphase mixtures were vigorously stirred by a method not reported. The similar results from resins of widely different cross-linking and porosity, both macroporous and non-macroporous, suggest that the rates were controlled mainly by intrinsic reactivity at the active site.

Effects of polymer structure on reaction of phenylacetonitrile with excess 1-bromobutane and 50% NaOH have been studied under conditions of constant particle size and 500 rpm stirring to prevent mass transfer limitations [103]. All experiments used benzyltrimethylammonium ion catalysts 2 and addition of phenylacetonitrile before addition of 1-bromobutane as described earlier. With 16–17% RS the rate constant with a 10% CL polymer was 0.033 times that with a 2% CL polymer. Comparisons of 2% CL catalysts with different % RS and Amberlyst macroporous ion exchange resins are in Table 6. The catalysts with at least 40% RS were more active that with 16% RS, opposite to the relative activities in most nucleophilic displacement reactions. If the macroporous ion exchange resins were available in small particle sizes, they might be the most active catalysts available for alkylation of phenylacetonitrile.

Table 6. Effect of % Ring Substitution on Activity for Alkylation of Phenylacetonitrile [103]

Catalyst[a]	% RS	av. diameter μm	$k \times 10^5$ s^{-1} mequiv^{-1}
2% cross-linked	16	223	57.8
2% cross-linked	50	229	171
Amberlyst A-27	40	466	82.2
Amberlyst A-27	40	411	105
Amberlyst A-26	90	502	49.5

[a] Amberlyst catalysts are macroporous from Rohm and Haas Co

No systematic study is available on other parameters in triphase alkylation of phenylacetonitrile, but the following isolated observations may be significant. Both dilution of the organic phase with benzene or cyclohexane and use of 10% NaOH in place of 50% NaOH greatly reduced the rate [101]. The benzyltrimethylammonium ion is attacked by hydroxide ion under the conditions of phenylacetonitrile alkylation. Repeated use of either Dowex ion exchange resins [101] or 2% CL, 16–50% RS resins [103] gave reduced activity.

Polymer-supported insoluble benzyltrimethylammonium ions [101,103] show about the same selectivity as soluble benzyltriethylammonium ion [99] for monoalkylation (Eq. (7)) over dialkylation (Eq. (8)) of phenylacetonitrile. Both gave about 85% monobutyl and 5% dibutyl products if stopped at the right time. Longer reaction times gave substantially more dibutyl product.

$$C_6H_5CH(n\text{-}C_4H_9)CN + n\text{-}C_4H_9Br + NaOH \rightarrow$$

$$C_6H_5C(n\text{-}C_4H_9)_2CN + NaBr + H_2O \tag{8}$$

3.3 Alkylation of Phenoxide and 2-Naphthoxide Ions

Alkylation of 2-naphthoxide ion (Eq. (6)) occurs mainly on carbon in aqueous solvents and mainly on oxygen in aprotic solvents. The product distribution is often used as a probe of the solvent environment in heterogeneous reactions. Brown and Jenkins [54] found that 40–100 % RS spacer chain catalysts *15* and *16* gave up to 98 % O-benzylation of 2-naphthoxide ion with benzyl bromide. The shorter spacer chain catalyst *16* gave 85 % O-alkylation, and a conventional benzyltrimethylammonium ion resin *2* gave about 70 % O-alkylation. Because of low activity, product distribution data were obtained with varied amounts of catalyst and were extrapolated to equimolar amounts of catalyst and substrate to obtain the catalyzed O/C product ratios. Interpretation of the data also was complicated by independent evidence that catalysts *15* adsorbed 2-naphthoxide ion, in addition to that bound by ion exchange [54]. Essentially the same results were obtained with catalysts *24* which lack the ester link in the spacer chain [106].

$$\text{(P)}-\text{(O)}-CH_2S(CH_2)_{11}\overset{\oplus}{N}(CH_3)_2R \quad I^{\ominus} \qquad 24: R = CH_3, \ CH_2CH_2OH$$

More recent investigations of the alkylation of 2-naphthoxide ion with benzyl bromide have led to conflicting interpretations of the active site environment in polystyrene-supported onium ion catalysts. Representative data from three laboratories are collected in Table 7. Montanari [107] attempted to duplicate the results of Regen [78] obtained in quiet toluene solutions of benzyl bromide (expts 1–6 of Table 7). With slightly different relative amounts and % RS of the catalysts, which should have minor effects on the results, the data are in complete disagreement between the two laboratories. The same reactions under rapidly stirred (1300 rpm magnetic) conditions proceded as fast in the absence as in the presence of the catalysts [107]. Both research groups reported a slow reaction favoring C-alkylation in quiet liquid/liquid mixtures with no solid catalyst [78, 107]. With no catalyst and rapid stirring only 16 % O-alkylation was detected [107]. In expts 4 and 6 of Table 7 with rapid stirring 62–72 % O-alkylation was detected, even though there was no increase in overall rate of reaction compared with that observed in the absence of polymer beads [107]. With stirring Brown and Henkins [54] also found as little as 16 % O-alkylation using 10 mol % of ion exchange resin *15* as catalyst. There is no way at present to reconcile the results of expts 1–6 in Table 7.

The data in Table 7 obtained with equimolar amounts of the polymeric catalysts and the 2-naphthoxide ion should be more reliable because all of the reactive anion is contained within the polymer. These conditions (expts 7–9) gave 100 % O-alkylation, indicating that the active site environment of the polystyrene-bound tri-n-butylphosphonium ion/naphthoxide ion pair or aggregate is aprotic even with the 60%RS polymer. However, the common benzyltrimethylammonium ion found in commercial ion exchange resins is more hydrophilic, giving both C- and O-alkylation (expts 10 and 11 of Table 7).

Benzylation of the phenoxide ion does not suffer from the complications of a rapid uncatalyzed reaction [107]. With 1300 rpm magnetic stirring to overcome mass transfer

Table 7. Benzylation of 2-Naphthoxide Ion

Expt.	Catalyst	mequiv catalyst	ml solvent	mmol $C_6H_5CH_2Br$	ml water	mmol $C_{10}H_7ONa$	temp, °C	mixing method	% conv.	% O-alkyl	Ref.
1	*1*, 17% RS, 1% CL	0.065	2.5	0.50	3.0	1.5	50	quiet	69	94	78)
2	*1*, 52% RS, 1% CL	0.125	2.5	0.50	3.0	1.5	50	quiet	36	19	78)
3	*1*, 10.6% RS, 1% CL	0.02	1.0	0.20	1.0	0.20	55	quiet	34	75	107)
4	*1*, 10.6% RS, 1% CL	0.02	1.0	0.20	1.0	0.20	55	1300 rpm	*a*	72	107)
5	*1*, 60% RS, 1% CL	0.02	1.0	0.20	1.0	0.20	55	quiet	32	75	107)
6	*1*, 60% RS, 1% CL	0.02	1.0	0.20	1.0	0.20	55	1300 rpm	*a*	62	107)
7	*1*, 10.6% RS, 1% CL	0.20	1.0	0.20	1.0	0.20	55	quiet	50	100	107)
8	*1*, 60% RS, 1% CL	0.20	1.0	0.20	1.0	0.20	55	quiet	41	100	107)
9	*5*, 25% RS, 2% CL	0.20	1.0	0.20	1.0	0.20	55	quiet	*a*	100	107)
10	*2*(Dowex-1)	0.20	1.0	0.20	1.0	0.20	55	quiet	*a*	56	107)
11	*2*(Dowex-1)	0.126	none	0.126	1.0	0.50	20	stirred	*a*	70	54)
12	*15*, 98% RS, 2% CL	0.126	none	0.126	1.0	0.50	20	stirred	*a*	98	54)

a Not reported

limitations the same tributylphosphonium ion catalysts as in expts 7–9 of Table 7 gave 100% O-alkylation, and the more hydrophilic benzyltrimethylammonium ion catalyst 2 gave 91.4% O-alkylation (Eq. (9)). [107]

$$
\begin{array}{c}
O^{\ominus} \\
\bigcirc
\end{array}
+ \ C_6H_5CH_2Br \longrightarrow
\begin{array}{c}
CH_2OC_6H_5 \\
\bigcirc
\end{array}
+ \
\begin{array}{c}
OH \\
\bigcirc \\
CH_2C_6H_5
\end{array}
\qquad (9)
$$

A slow non-competing liquid/liquid reaction with no catalyst present gave only 78% O-alkylation. Thus the active site of the lipophilic phosphonium ion catalysts appears to be aprotic, just as in analogous phase transfer catalyzed alkylations with soluble quaternary ammonium salts [60]. Regen [78] argued that the onium ion sites of both the 17% and the 52% RS tri-n-butylphosphonium ion catalysts 1 are hydrated, on the basis of measurements of water contents of the resins in chloride form. Montanari has reported measurements that showed only 3.0–3.8 mols of water per chloride ion in similar 25% RS catalysts [74]. He argued that such small hydration levels do not constitute an aqueous environment for the displacement reactions. No measurements of the water content of catalysts containing phenoxide or 2-naphthoxide ions have been reported.

3.4 Macroporous and Isoporous Polymer Supports

Macroporous and isoporous polystyrene supports have been used for onium ion catalysts in attempts to overcome intraparticle diffusional limitations on catalyst activity. A macroporous polymer may be defined as one which retains significant porosity in the dry state [68–71]. The terms macroporous and macroreticular are synonomous in this review. Macroreticular is the term used by the Rohm and Haas Company to describe macroporous ion exchange resins and adsorbents [108]. The terms microporous and gel have been used for cross-linked polymers which have no macropores. Both terms can be confusing. The micropores are the solvent-filled spaces between polymer chains in a swollen network. They have dimensions of one or a few molecular diameters. When swollen by solvent a macroporous polymer has both solvent-filled macropores and micropores created by the solvent within the network. A gel is defined as a solvent-swollen polymer network. It is a macroscopic solid, since it does not flow, and a microscopic liquid, since the solvent molecules and polymer chains are mobile within the network. Thus a solvent-swollen macroporous polymer is also microporous and is a gel. Non-macroporous is a better term for the polymers usually called microporous or gels. A sample of 200/400 mesh spherical non-macroporous polystyrene beads has a surface area of about 0.1 m²/g. Macroporous polystyrenes can have surface areas up to 1000 m²/g.

Macropores are created during synthesis by polymerization in the presence of a non-polymerizable diluent from which the polymer precipitates as it cross-links [68–71]. After complete polymerization the polymer contains diluent-filled pores.

The diluent is removed by distillation or by washing and drying. Three different classes of diluents have been used: non-solvents, good solvents, and oligomers having the same basic structure as the network [70]. A good solvent does not dissolve, but swells a cross-linked network. Three major factors influence whether the polymer will retain macroporosity upon drying: a) the degree and chemical nature of cross-linking, b) the diluent used in the synthesis, and c) the solvent from which the polymer is dried. All three types of diluents lead to macroporous polymers when the network is highly cross-linked. The higher the divinylbenzene content used with a given diluent, the greater the surface area and the smaller the average pore size [68-70]. Macroporous polystyrenes have been reported with average pore sizes ranging from <50 Å to $>50,000$ Å. The poorer a solvent the diluent is for the polymer, the more likely the polymer is to retain macroporosity upon drying [109]. A polystyrene synthesized with heptane as diluent will retain macroporosity at a much lower level of cross-linking than a polystyrene synthesized with toluene as diluent. With nonsolvent diluents it is possible to synthesize macroporous polystyrenes cross-linked with as little as 1 % divinylbenzene. The commonly held impression that macroporous polymers must be highly cross-linked applies only to networks synthesized in the presence of good solvents or dried from good solvents. The total pore volume of the polymer is smaller when it is dried from a good solvent than when it is dried from a poor solvent [109]. A cross-linked polystyrene is most likely to retain macroporosity if it is dried from methanol or hexane and least likely to retain macroporosity if it is dried from benzene or chloroform. The three classes of diluents often lead to quite different porosities and morphologies [70].

The theoretical advantages of a macroporous polymer as a catalyst support lie in its high internal surface area and in the small size of the microparticles which are fused together to form the macroparticle. If a catalyzed reaction takes place on the internal surface, reactant molecules must undergo mass transfer to the external surface, diffusion through the quiet liquid-filled macropores, and chemical reaction on the internal surface. If a catalyzed reaction takes place within the gel phase, the diffusion path of the reactant must continue from the internal surface into the polymer network, but the intraparticle diffusion distance is much shorter than in a non-macroporous polymer of the same macroscopic particle size. Thus a macroporous catalyst will be more active than a non-macroporous catalyst whenever the reaction rate is limited in part by intraparticle diffusion, and the diffusion of reactants through the liquid-filled macropores is faster than diffusion through the swollen polymer network [64, 72]. In principle, macroporosity should not affect mass transfer rates because mass transfer depends only upon the external surface area of the particle [61]. Macroporosity also should not affect the rate of intraparticle diffusion as long as the same amount and type of cross-linking is present the macroporous polymer as in the comparable non-maroporous polymer. Only the length of the diffusion path is different. Of course, if one compares a highly cross-linked macroporous polymer with a lightly cross-linked non-macroporous polymer, the rate of diffusion through the network will be greater in the less cross-linked polymer.

The first attempts to use macroporous polystyrene supports for onium ion catalysts were reported in the early papers of Brown and Jenkins [54] and of Regen [89]. The lightly cross-linked macroporous Rohm and Haas XE-305, 46 % RS as spacer chain catalyst *15*, gave O- and C-alkyl products from 2-naphthoxide ion and benzyl bromide

almost identical to those obtained with a 98 % RS, 2 % CL non-macroporous analog [54].
The macroporous catalyst was slightly more active, although no kinetic data were
reported.

Attempts to improve the activity of polystyrene-bound phosphonium ions *1* by
use of macroporous supports have been disappointing. Chloromethylation and tri-
n-butylphosphonium salt formation with the higly porous adsorbents XAD-2 and
XAD-4 (surface areas ca. 300 and 750 m^2/g respectively [110]) gave catalysts much less
active than the standard 2 % CL non-macroporous catalyst for reaction of chloride ion
with n-decyl methanesulfonate (Eq. (5)) [77]. The less highly cross-linked macroporous
catalysts reported in the same paper were not good tests of macroporous supports
because they had surface areas of <1 m^2/g [77]. A series of macroporous polystyrene
phosphonium ion catalysts *1* with 2, 4, 6, 10, 20, 50, and 75 % divinylbenzene cross-
linking and high surface areas was tested for reaction of cyanide ion with 1-bromo-
octane in toluene (Eq. (3)) [86]. Activity decreased as cross-linking increased throughout
the series, and the lightly cross-linked macroporous catalysts were slightly less active
than the analogous non-macroporous catalysts with the same % divinylbenzene
(Fig. 9). Determinations of the cyanide and bromide ion contents of catalysts recovered
from reaction mixtures showed that both the macroporous and the non-macroporous
catalysts contained more bromide ion than cyanide ion. Approximate calculations
based on the reaction kinetics and the fractional attainment of ion exchange equilibrium gave diffusion coefficients of bromide and cyanide ion in the catalysts in the
range 10^{-8} to 10^{-9} cm^2/s for 2 % and 10 % CL catalysts at 90 °C. The diffusivities
were much smaller than any previously reported for ion exchange in aqueous media [63].
The toluene phase must retard ion exchange. Normally in aqueous systems macroporosity speeds ion exchange in highly corss-linked resins by enabling diffusion to
occur mostly through the macropores. We have concluded tentatively that in triphase
catalysis the macropores were filled with the organic phase which slowed greatly the
ion transport through the macropores. The internal surface of the polystyrene-bound

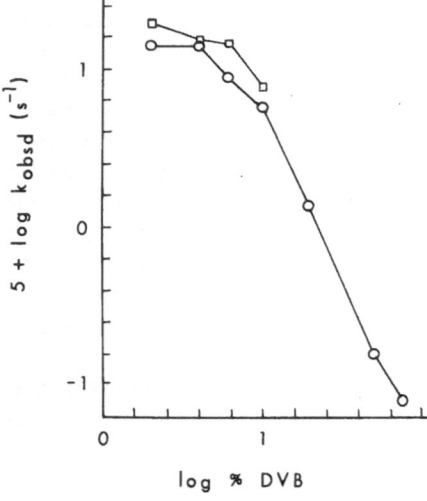

Fig. 9. Dependence of k_{obsd} on % CL for non-macroporous (□) and macroporous (○) catalysts for
reaction of 1-bromooctane in toluene with aqueous
NaCN at 90 °C. (Reprinted with permission from
Ref. [86]. Copyright 1982 American Chemical Society)

catalysts were lipophilic and adsorbed toluene in preference to water. To overcome the ion exchange limitations of macroporous catalysts in triphase mixtures a more hydrophilic polymer surface may be needed.

Isoporous polymers resemble macroporous polymers but are prepared by cross-linking a linear or very lightly cross-linked polymer in a good solvent with a difunctional alkylating agent and a Friedel-Crafts catalyst [71]. Tundo [112] prepared a tri-n-butylphosphonium ion catalyst 25 from soluble polystyrene by alkylation and phosphination as shown in Scheme 3. The catalyst 25 was highly swellable and contained 1.27 mequiv of bromide ion per g. As a finely ground powder it was highly active for reactions of 1-bromooctane with iodide, cyanide, and phenoxide ions and for alkylation of 1-phenyl-2-propanone with 1-bromobutane and aqueous 50% NaOH. In a subsequent kinetic study of the reaction of 1-bromooctane with iodide ion, it was the most active polymer-bound insoluble catalyst [74].

Scheme 3

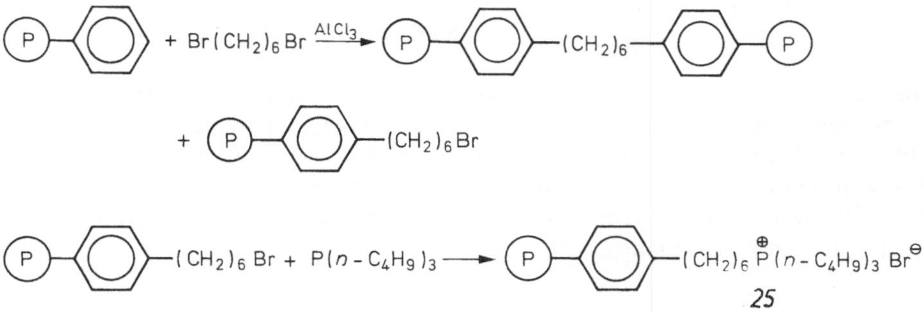

By a method similar to isoporous polystyrene synthesis, poly(4-vinylpyridine) has been cross-linked with dibromoalkanes to give insoluble networks containing pyridinium ions (26) [113].

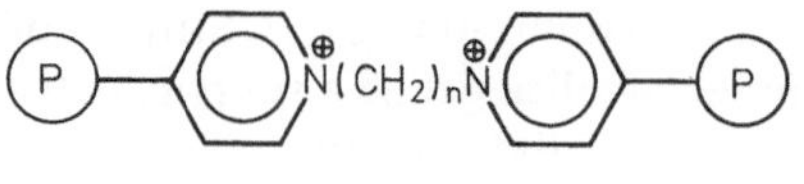

$26: n = 2, 3, 4, 6, 8, 10, 12$

Particle sizes and swelling of the polymers were not reported. Catalytic activity for reaction of 1-bromooctane with cyanide ion (Eq. (3)) decreased with the length of the cross-link from $n = 2$ to $n = 3$ and then increased up to $n = 12$. The reaction mixtures were analyzed only after 24 h and 48 h, and 900 rpm magnetic stirring was used, so the activity differences were probably due to intrinsic reactivity. The catalysts with longer cross-linking chains were more active probably because they provided more lipophilic environments for the reactions. The higher activity of the $n = 2$ catalyst was probably due to incomplete quaternization, leaving effectively a lower % RS polymer. The yield of the cross-linked polymer with $n = 2$ was about 40% compared with >80% for all of the other chain lengths [113]. Isoporous polymers appear promising as catalyst supports, but it is not possible to compare them well with either conventional non-macroporous or macroporous polymers until careful experiments have been carried out on their swelling properties (as a method to compare their cross-linking with other polystyrene networks) and on mixing and particle size effects on the activities of the derived catalysts.

3.5 Silica Gel and Alumina Supports

Quaternary onium ions bound to silica gel were reported as phase transfer catalysts initially by Tundo [114] and by Rolla and co-workers [115]. The short spacer chain catalyst 27 (1.0 mmol/g) was more active than the longer spacer chain phosphonium ion catalyst 28

$$\text{(SiO}_2\text{)}\!-\!Si(CH_2)_3\overset{\oplus}{P}(n\text{-}C_4H_9)_3 \qquad\qquad \text{(SiO}_2\text{)}\!-\!Si(CH_2)_3NHCO(CH_2)_{10}\overset{\oplus}{Z}(n\text{-}C_4H_9)_3$$
$$27 \qquad\qquad\qquad\qquad\qquad\qquad 28 : Z = P, N$$

$$\text{(SiO}_2\text{)}\!-\!Si(CH_2)_3\overset{\oplus}{N}R_3\ \ X^{\ominus}\quad 29 : R = CH_3, C_2H_5, n\text{-}C_4H_9, n\text{-}C_8H_{17}, n\text{-}C_{16}H_{33}$$
$$X = Cl, Br, I$$

for reaction of potassium iodide with 1-bromooctane, sodium borohydride reduction of acetophenone in benzene, and reaction of potassium phthalimide with benzyl chloride [114]. The supported catalysts were less active for the displacement reactions, but were more active for the borohydride reduction than soluble onium ion catalysts [114]. As with polystyrene-supported and soluble catalysts, the more lipophilic quaternary ammonium ions 29–31

$$\text{(SiO}_2\text{)}\!-\!SiC_6H_4CH_2\overset{\oplus}{Z}(n\text{-}C_4H_9)_3\ Cl^{\ominus} \qquad \text{(SiO}_2\text{)}\!-\!Si(CH_2)_6\overset{\oplus}{N}(n\text{-}C_4H_9)_3\ Br^{\ominus}$$
$$30 : Z = N, P \qquad\qquad\qquad\qquad\qquad 31$$

(0.06–0.33 mmol/g) bound to silica gel were much more active than the trimethyl-ammonium ion 29 (R = CH$_3$, 0.50 mmol/g) for the displacement of bromide from 1-bromohexane in toluene by aqueous potassium iodide [115]. Stirring of the reaction mixtures was vital [115]. The disadvantage of silica gel as a support is that it dissolves in aqueous alkali. It is unsuitable for any of the common phase transfer catalyzed reactions that require strongly alkaline conditions. Even concentrated aqueous sodium cyanide destroyed the support [115].

Phosphonium ion catalysts 27 and 28 have been investigated in detail in the reaction of 1-bromooctane with iodide ion [116–118]. Magnetic stirring of at least 600 rpm was required to achieve maximum rates. With no stirring there was a rapid initial reaction due to reactants already within the catalyst followed by a much slower mass transfer limited reaction [117]. Ion exchange experiments with iodide-131 indicated fast penetration of the catalyst by iodide under fast stirring conditions. In the stirred reaction of n-octyl methanesulfonate with aqueous potassium bromide, catalysts recovered after 15–23% conversion contained 94–100% bromide ion at the active sites, but without stirring phosphonium ion catalysts 27 and 28 contained 28% and 68% bromide ion respectively, with the remaining ion exchange sites occupied presumably by methanesulfonate ion [117].

No experiments with variation in particle size of the silica gel have been done to study intraparticle diffusion effects. In silica gel such diffusion would be only through the pores (analogous to the macropores of a polystyrene) since the active sites lie on the internal surface. The silica gel used by Tundo had a surface area of 500 m^2/g and average pore diameter of 60 Å. [116]. Phosphonium ion catalyst 28 gave rates of iodide displacements that decreased as the 1-bromoalkane chain length increased from C_4 to C_8 to C_{16}. The selectivity of 28 was slightly less than that observed with soluble catalyst hexadecyltri-n-butylphosphonium bromide [118]. Consequently the selectivity cannot be attributed to intraparticle diffusional limitations.

Table 8. Activities of Silica Gel and Alumina-Supported Catalysts for the Reaction of 1-Bromobutane with Aqueous KI [118]

Catalyst[a]	mmol/g	$10^4 k_{obsd}$, s^{-1} [b]
29:R = n-C_4H_9	0.16	1.18
27	0.34	1.88
28:Z = N	0.43	1.36
28:Z = P	0.52	1.74
32:Z = N	0.11	1.36
32:Z = P	0.31	3.10
33:Z = N	0.14	2.51
33:Z = P	0.25	3.68
n-$C_{16}H_{33}\overset{+}{N}$(n-$C_4H_9$)$_3Br^-$	—	9.21
n-$C_{16}H_{33}\overset{+}{P}$(n-$C_4H_9$)$_3Br^-$	—	7.14

[a] 0.5 mol % based on C_4H_9Br;
[b] 10.0 mmol C_4H_9Br, 30.0 mmol KI, 3.25 ml H_2O, 80 °C, 1000 rpm magnetic stirring

Activities of tri-n-butylammonium and tri-n-butylphosphonium ions with two different spacer chain lengths are compared in Table 8 [118]. The greater activity of the phosphonium ions is opposite to what has been reported for analogous soluble phase transfer catalysts [119]. Activities of the catalysts bound to silica gel were as high as activities of soluble catalysts adsorbed to silica gel [118]. Without some independent determination of the role of intraparticle diffusion it is not possible to determine whether the reduced activity of the adsorbed catalysts is due to lower intrinsic activity at the silica gel surface or to diffusional limitations. The size selectivity for alkyl bromides suggests that intraparticle diffusion was not a problem.

Variation of the organic solvent in the 1-bromooctane reaction with iodide ion catalyzed by 27 and 28 (Z = P) showed rate constants decreasing in the order heptane > chlorobenzene > toluene [116]. The higher activity in heptane was attributed to greater adsorption of the 1-bromooctane onto the silica gel. Rates of reduction of ketones by sodium borohydride in cyclohexane, benzene, and chlorobenzene using the same catalysts correlated approximately with the adsorption equilibrium constants for the ketones from the solvents onto the silica gel (cyclohexane > chlorobenzene > benzene) [116].

As ion exchange resins the silica gel catalysts bind oxy anions more strongly than chloride ion [119]. Relative binding constants to ion exchange resin 29 (X = n-butyl) are HCO_3^-, HSO_4^-, $C_6H_5O^- > F^- > CH_3CO_2^- > NO_3^- > I^- > Br^- > Cl^-$, wheras on Dowex-1, a typical polystyrene-based benzyltrimethylammonium ion resin 2, the order is $I^- > C_6H_5O^- > HSO_4^- > NO_3^- > Br^- > Cl^- > HCO_3^- > CH_3CO_2^- > F^-$ [119]. The binding constants indicate aqueous environments of the active sites in the silica gel catalysts [119]. The active site environment of 28 (Z = P) is more aqueous than that of the lipophilic polystyrene-bound catalysts but less aqueous than that of Dowex-1 according to 85/15 O/C benzylation of 2-naphthoxide ion under stoichiometric reaction conditions [107]. (Compare with Table 7).

Alumina also has been used as a support for tri-n-butylammonium and tri-n-butylphosphonium ion catalysts 32 and 33, prepared by the same method used with silica gel [118].

$$\boxed{Al_2O_3} - Si(CH_2)_3 \overset{\oplus}{Z} (n\text{-}C_4H_9)_3 \ Br^{\ominus}$$
$$32 : Z = N, P$$

$$\boxed{Al_2O_3} - Si(CH_2)_3 NHCO(CH_2)_{10} \overset{\oplus}{Z} (n\text{-}C_4H_9)_3 \ Br^{\ominus}$$
$$33 : Z = N, P$$

The alumina-bound catalysts were more active than those on silica gel by factors of 2.1 or less, and were less active than the corresponding soluble catalysts for reaction of 1-bromooctane with iodide ion (Table 8). Reaction mixtures were stirred vigorously to overcome mass transfer limitations. Recovered catalyst 32 (Z = P) after 37% conversion contained 0.25 mequiv of iodide and 0.06 mequiv of bromide per gram when the triphase reaction was conducted with vigorous stirring and 0.14 and 0.16 mequiv per gram with no stirring. Therefore, mass transfer of the ions was a rate limiting factor with no stirring but may not have been rate limiting with stirring. Particle sizes were not studied. Most experiments were done with alumina which had a surface area of 180–200 m²/g and an average pore diameter of 60 Å. A few experiments used 70 m²/g, 150 Å alumina. The larger pore alumina was more active with 1-bromobutane and with 1-bromooctane but less active with 1-bromohexadecane. Why the larger pore alumina was more size-selective is unknown. Complete pore size distributions of the aluminas might provide an explanation. The alumina catalysts were much more size-selective than the silica gel catalysts [118], indicating much greater intrapore diffusional control in alumina. Since the alumina catalysts also gave higher observed rates, the alumina catalysts must have higher intrinsic activities relative to the silica gel catalysts than shown simply by the relative rate constants.

Both alumina and silica gel are more stable physically than the common polystyrene supports. The alumina-bound catalysts are particularly promising because of their higher activity and higher selectivity compared with the silica gel-bound catalysts. Alumina also is stable in alkali. The alumina-bound catalysts 32 and 33 worked well for reaction of 1-bromooctane with concentrated aqueous sodium cyanide [118].

3.6 Continuous Flow Reactors

Triphase liquid/liquid/solid catalysis has been carried out with a continuous flow reactor [120]. A mixture of 1-bromooctane in o-dichlorobenzene and aqueous potassium iodide was pumped upward through a bed of polystyrene-supported phosphonium ion catalyst (not identified but obtained from Montanari and Tundo). The liquid phases were mixed with a magnetic stirring bar at the entrance to the packed tube. Up to 90% conversion was attained in 8 h at 80 °C by continuous recycling of 100 g of the liquids through 0.5 g of the catalyst. The overall rate of conversion was slower than in a stirred batch reactor using the same amounts of reactants and catalyst, which suggests that mass transfer was less efficient in the fixed bed reactor. It is easy to imagine that the two liquids might have channeled through the catalyst bed unless flow within the fixed bed was turbulent.

Triphase catalysis has also been reported in flow systems in which the organic reactant in the vapor phase passed over a bed of a phosphonium ion phase transfer catalyst and a solid inorganic salt at 150 °C [121-123]. Quaternary ammonium ions were not stable enough for use at such high temperature. In a typical experiment a slurry of potassium iodide, a soluble or silica gel-bound phosphonium salt, and silica gel in methanol/water was dried and placed in the tubular reactor. Vapor of 1-bromooctane flowed in and a mixture of 1-bromooctane and 1-iodooctane was collected. Conversion was 78% after one pass and 92% after three passes [121]. The best catalysts were those which melted below the reaction temperature, forming a film containing quaternary phosphonium ions on the solid inorganic salt [123]. The reactions most likely took place in the film.

3.7 Synthetic Applications

A variety of synthetic applications of polystyrene-supported quaternary ammonium ions was reported by Regen [124]. He was the first to demonstrate the value of low %RS for reactions of lipophilic organic compounds. Many other synthetic applications have been reported, and the reader is directed to other reviews for them [16,39,42,43].

One intriguing synthetic possibility is the use of insoluble chiral polymeric quaternary ammonium ions as catalysts for asymmetric syntheses. This has been tried with no great success as of the time of writing of this review. The potential advantages of insoluble chiral catalysts are a) they are catalysts, not reagents, and are required only in small amounts, and b) they allow recovery and reuse of the expensive chiral species. The only high optical yield experiment using a chiral insoluble polymer as a catalyst (of which we are aware) is epoxidation of chalcone in 85% chemical and 93% optical yield in toluene with basic aqueous hydrogen peroxide and an insoluble 10-mer of (S)-alanine as catalyst [125]. Reuse of the polymer was reported in only one case which gave reduced optical yield [125]. Numerous other polymers and reactions have given < 50% enantiomeric excesses [16,126-138].

Commercial anion exchange resins have been used extensively for syntheses. A stoichiometric quantity of the anionic reactant is bound to the quaternary ammonium ion resin and allowed to react with an organic substrate in an organic solvent. The resin can be recovered and reused. Since the resins are not catalysts, they are not discussed further here. See other reviews for many examples [16,39,40,41,139].

4 Triphase Catalysis by Crown Ethers and Cryptands

Polymer-supported crown ethers and cryptands were found to catalyze liquid-liquid phase transfer reactions in 1976 [55]. Several reports have been published on the synthesis and catalytic activity of polymer-supported multidentate macrocycles. However, few studies on mechanisms of catalysis by polymer-supported macrocycles have been carried out, and all of the experimental parameters that affect catalytic activity under triphase conditions are not known at this time. Polymer-supported macrocycle catalysts have been applied for nucleophilic displacements and for a few other reactions [55,93,140–148].

4.1 Mass Transfer

The dependence of k_{obsd} on stirring speed for Br-I exchange reactions with polymer-supported crown ethers 34 and 35 has been determined under the same conditions as with polymer-supported phosphonium salts 1 and 4 [149]. Reaction conditions were 90 °C, 0.02 molar equiv of 100–200 mesh catalyst, 16–17% RS, 2% CL, 20 mmol of 1-bromooctane, 200 mmol of KI, 20 ml of toluene, and 30 ml of water. Reaction rates with 34 and 35 increased with increased stirring speed up to 400 rpm, and were constant above that value. This result resembles that with polymer-supported onium ion catalysts and indicates that mass transfer as a limiting factor can be removed in experiments carried out at stirring speeds of 500–600 rpm, whatever kind of polymer-supported phase transfer catalyst is used.

4.2 Intraparticle Diffusion

The contribution of intraparticle diffusion to rate limitation can be seen from dependence of k_{obsd} on the particle size of catalysts. Fig. 10 shows the effect of particle size on k_{obsd} for iodide displacement reactions (Eq. (4)) with catalysts *34*, *35*, and *41* [149].

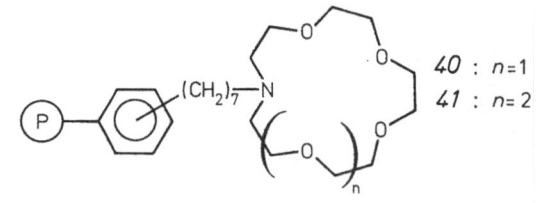

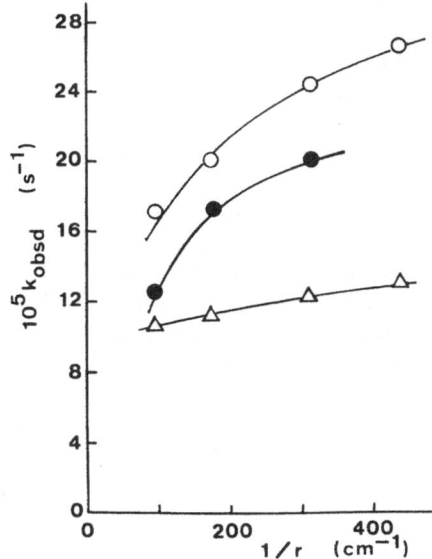

Fig. 10. Dependence of k_{obsd} on mean particle size of 2% CL catalyst for reaction of 1-bromooctane in toluene with 10 molar equiv of KI in water and 0.02 molar equiv of catalyst at 90 °C; 16% RS catalyst *34* (△), 17% RS catalyst *35* (●), 14% RS catalyst *41* (○); stirring speed 550–600 rpm

Rates with *34* depended slightly on the particle size, whereas rates with *35* and *41* depended significantly on particle size. These results indicate that reactivity with *34* is limited mainly by intrinsic reactivity, and reactivity with *35* and *41* is limited by a combination of intraparticle diffusion and intrinsic reactivity. Such $1/r$ dependences of k_{obsd} are similar to those with polymer-supported benzyltrimethylammonium ion *2* and benzyltri-n-butylphosphonium ion *1*. (See Fig. 4 and Table 1 in Sect. 3.1.2).

As with polymer-supported onium ions the degree of cross-linking of the polymer support is likely to affect mainly intraparticle diffusion in reactions with polymer-supported crown ethers or cryptands. The activity of catalyst *37* decreased by a factor of about 3 as % CL with divinylbenzene changed from 1% to 4.5% [146].

The introduction of a spacer chain between the polymer backbone and the active site is expected to facilitate intraparticle diffusion. Increased swelling power of sol-

Table 9. Influence of Solvent and of Spacer Chain Length on Activity for Reaction of 1-Bromooctane with Iodide Ion at 90 °C[146]

Catalyst, 2% CL	mequiv/g	% RS	$10^5 k_{obsd}$ (s^{-1})		
			n-Heptane	Toluene	Chloro-benzene
37	1.0	23	6	22	21
38	0.48	14	8	26	26
39	0.48	14	19	28	27
36	0.26	3.6	11	43	45

vents also improves intraparticle diffusion. Table 9 shows effects of solvents and spacer chain length on k_{obsd} for halide exchange reactions [146]. The rate with catalyst *37* having an 11-atom spacer was smaller in n-heptane than in toluene or chlorobenzene, because the heptane hardly swelled the catalyst. Increased spacer chain length (catalyst *39* with a 35-atom spacer) led to easier intraparticle diffusion even in a poor solvent, and the rate approached that obtained in a good solvent. However, the contribution of solvents and spacer chains to intrinsic reactivity must also be considered.

The activity of polymer-supported crown ethers depends upon the degree of substitution of the polymer support. Fig. 11 reports dependence of k_{obsd} on % RS and solvent for iodide displacement reactions (Eq. (4)) with catalysts *34, 35* and *41* [149]. The rate with 6% RS *41* was smaller than that with 17% RS *35*, though the former catalyst had a 7-atom spacer. Reduced % RS makes the catalysts more lipophilic, and results in the slower intraparticle diffusion of the KI. Therefore, the lowest % RS catalysts

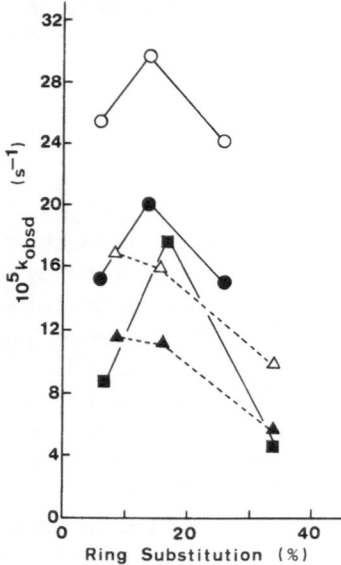

Fig. 11. Effects of % RS and solvent on k_{obsd}. Catalyst/solvent: *34*/toluene (▲), *34*/chlorobenzene (△), *35*/toluene (■), *41*/toluene (●), *41*/chlorobenzene (○). Experimental conditions are the same as in Fig. 10

35 and *41* show lower activity. Similar results have been obtained for reactions catalyzed by lower % RS spacer-modified phosphonium ion catalysts. (See Fig. 7 in Section 3.1.3). Higher activity of such low % RS catalysts does not appear unless the particle size is small, because the activity depends largely on the particle size [80, 149].

4.3 Intrinsic Reactivity

Under aqueous/organic two phase conditions cryptands with lipophilic groups extract alkali metal salts from the aqueous to the organic phase more effectively than crown ethers with lipophilic groups [150, 151]. Moreover, the reactivity of nucleophiles with cryptate counterions is higher (ca. 10–20 times) than that of nucleophiles with crown ether counterions under classical phase transfer conditions [150, 151]. In the complexes with cryptands cation-anion interactions are minimized, and the anion is highly active. In the complexes with crown ethers the anion can still interact with the complexed cation from a direction perpendicular to the plane of the ring, and the complexes exhibit less anion reactivity. In conclusion, cryptands are superior to crown ethers as phase transfer catalysts.

The activity of polymer-supported crown ethers is a function of % RS as shown in Fig. 11 [149]. Rates for Br-I exchange reactions with catalysts *34*, *35*, and *41* decreased as % RS increased from 14–17% to 26–34%. Increased % RS increases the hydrophilicity of the catalysts, and the more hydrated active sites are less reactive. Less contribution of intraparticle diffusion to rate limitation was indicated by less particle size dependence of k_{obsd} with the higher % RS catalysts [149].

Complexation constants of crown ethers and cryptands for alkali metal salts depend on the cavity sizes of the macrocycles [152, 153]. In phase transfer nucleophilic reactions catalyzed by polymer-supported crown ethers and cryptands, rates may vary with the alkali cation. When a catalyst *41* with an 18-membered ring was used for Br-I exchange reactions, rates decreased with a change in salt from KI to NaI, whereas catalyst *40* bearing a 15-membered ring gave the opposite effect (Table 10) [149]. A similar rate difference was observed for cyanide displacement reactions with polymer-supported cryptands in which the size of the cavity was varied [141]. Polymer-supported phosphonium salt *4*, as expected, gave no cation dependence of rates (Table 10).

Table 10. Activities of Polymer-bound Crown Ether and Phosphonium Salts for Reactions of 1-Bromooctane in Toluene with Iodide or Cyanide Ions at 90 °C[149]

Catalyst, 2% CL 100/200 mesh		mequiv/g	% RS	Nucleophile	$10^5 \, k_{obsd}$ (s^{-1})
Crown	*41*	0.82	14	KI	20.1
	41	0.82	14	NaI	11.4
	41	0.82	14	KCN	1.5
	40	0.71	11	KI	5.0
	40	0.71	11	NaI	12.4
P$^+$	*4*	0.99	17	KI	28.5
	4	0.99	17	NaI	28.0
	4	0.99	17	NaCN	49.8[79]

Under aqueous/organic two phase conditions the amount of inorganic salt extracted by crown ethers into the organic phase and the reaction rates depend strongly on both the cation and the anion [150,151]. Polymer-supported crown ethers, as well as soluble crown ethers, are less effective catalysts for liquid-liquid phase transfer reactions using highly electronegative, slightly polarizable (hard) anions such as cyanide than they are for the reactions of polarizable (soft) anions such as iodide. In practice *40* and *41* are slighly less active than polymer-supported phosphonium catalyst *4* (Table 10) [149]. A similar behavior of catalyst *42* has been observed for cyanide and iodide displacement reactions [145].

42   *43*

However, with catalyst *37* having a slightly longer spacer, the difference in rates for cyanide and iodide displacements was smaller, than that with catalysts *41* and *42*. Catalyst *43* having the crown moiety as an integral part of the polymer backbone showed higher activity for cyanide displacement than for iodide displacement [145]. Rates for both displacement reactions with cryptand *36* were larger than those with crown *37*, and the difference in rates for the cyanide and iodide reactions decreased further [146]. This must be due to the large complexation constants of cryptands and the high activity of cryptand complexes compared with crown ether complexes.

Spacer chains affect intrinsic reactivity as well as intraparticle diffusion. Rates for Br-I exchange reactions with spacer-modified catalyst *41* were larger than those with catalyst *35* containing no spacer (Fig. 11). An aliphatic spacer makes the catalyst more lipophilic and the intrinsic reactivity of the active site larger, though the intraparticle diffusity of an inorganic reagent is reduced. It is not known at this time how intrinsic reactivity contributes to the rate increase.

The activity of polymer-supported crown ethers depends on solvent. As shown in Fig. 11, rates for Br-I exchange reactions with catalysts *34* and *41* increased with a change in solvent from toluene to chlorobenzene. Since the reaction with catalyst *34* is limited substantially by intrinsic reactivity (Fig. 10), the rate increase must be due to an increase in intrinsic reactivity. The reaction with catalyst *41* is limited by both intrinsic reactivity and intraparticle diffusion (Fig. 10), and the rate increase from toluene to chlorobenzene corresponds with increases in both parameters. Solvent effects on rates with polymer-supported phase transfer catalysts differ from those with soluble phase transfer catalysts [60]. With the soluble catalysts rates increase (for a limited number of reactions) with decreased polarity of solvent [60], while with the polymeric catalysts rates increase with increased polarity of solvent [74]. Solvents swell polymer-supported catalysts and influence the microenvironment of active sites as well as intraparticle diffusion. The microenvironment, especially hydration state, of active sites of polymer-supported phase transfer catalysts may or may not be the same as that of soluble phase transfer catalysts. Systmatic investigations on the microenvironment are needed to account for the solvent effects.

Rates for Br-I exchange reactions were 1.5-fold higher with 10% RS, 1% CL catalyst *37* when the amount of KI was changed from 2.4 to 8.0 mmol in 0.75 ml of water [146]. Rates for the same reactions with 26–34% RS, 2% CL catalysts *35* and *41* hardly changed as the KI concentration was increased from 6.7 M to 10.0 M. Rates with 14–17% RS *35* and *41*, and with 7% RS *35*, increased by a factors of 1.5 and 2, respectively, with that increase in the KI concentration [149]. Apparently the concentration of inorganic salts in the aqueous phase affects complexation constants and/or intrinsic reactivity, especially the hydration state of the active site. The activity of lower % RS catalysts depends more on the salt concentration than does the activity of higher % RS catalysts, because the former are more lipophilic.

Polymer-supported onium ions are relatively unstable under severe conditions, especially concentrated alkali [154]. Polymer-supported crown ethers and cryptands are stable under such conditions. In practice, they could be reused without loss of catalytic activity for the alkylation of ketones under basic conditions, whereas the activity of polymer-supported ammonium ion *7* decreased by a factor of 3 after two recycles of the catalyst [147].

5 Triphase Catalysis by Polymer-Supported Cosolvents

At almost the same time as other polymer-supported phase transfer catalysts were first reported, polymer-supported solvents and cosolvents were found to be effective catalysts for phase transfer reactions [155,156]. Dipolar aprotic solvents such as hexamethylphosphoramide (HMPA) [157], dimethylsulfoxide (DMSO) [158], and tertiary amides [159,160] are well known to coordinate strongly with alkali and alkaline earth metal cations, and hence promote nucleophilic displacement reactions of the anions [161]. Catalysts *44* [155,162,163] and *45* [163],

$$\text{(P)}\!-\!\!\bigcirc\!\!-\!\text{CH}_2\text{N}\underset{\underset{R}{|}}{-}\underset{\underset{O}{\parallel}}{P}[\,\text{N(CH}_3)_2]_2 \qquad \begin{array}{l} \textit{44} : R = CH_3 \\ \textit{45} : R = H \end{array}$$

in which HMPA analogues were supported on non-macroporous and on macroporous polystyrene resins, were used successfully as catalysts for cyanide, iodide, hydroxide, and phenoxide [82] displacements and borohydride reductions under triphase conditions. The activity of 7% RS, 30% CL macroporous catalyst *45* was higher than that of 27% RS, 10% CL macroporous catalyst *44*. The difference was attributed to the presence of an amino hydrogen atom in *45*, though an exact comparison was difficult because the % CL and the % RS of the two catalysts were so different. The anions of inorganic reagents can interact with the acidic hydrogen atom of catalyst *45* by hydrogen bonding, and the phosphoryl oxygen atom interacts with the cations. Such a cooperative interaction may facilitate extraction of inorganic reagents into the polymer matrix. In practice, soluble catalysts corresponding to *45* were superior to those corresponding to *44* in both the ability to extract alkali metal salts and catalytic activity for liquid-liquid phase transfer reactions [164].

In the case of 1 % CL non-macroporous catalyst *44* the catalytic activity (shown in Ref. [162] on a weight basis and converted here to a molar basis) increased as the % RS increased from 17 % to 92 % for chloride displacements, decreased as % RS increased for hydroxide displacements, and hardly changed as % RS increased for cyanide displacements [162]. Rates of reactions catalyzed by polymer-supported HMPA analogues are conciderably less than those catalyzed by polymer-supported phosphonium salts. Therefore, reactions with catalysts *44* and *45* are likely to be limited not by intraparticle diffusion but by intrinsic reactivity. The overall catalytic activity is determined by two factors, the amount of inorganic salt extracted into the polymer matrix and the reactivity of the salt extracted. Increased loading leads to an increase in the hydrophilicity of catalysts, and hence to greater extraction of salts. However, increased hydrophilicity may reduce the anion reactivity of the extracted salts, particularly salts containing hard anions. Detailed studies of the hydrophile/lipophile balance of the catalysts are required to get more active polymer-supported cosolvent catalysts.

Recently, poly(N,N-dialkylacrylamide) *46* cross-linked with 1 mol % N,N'-dimethyl-N,N'-ethylenebis(acrylamide) was tested as a triphase catalyst [165].

$$\text{--}[\text{CH}_2\text{CH}]_n\text{--} \qquad 46 : R = CH_3 , C_2H_5 , \ n\text{-}C_3H_7 , \ n\text{-}C_4H_9 , \ n\text{-}C_8H_{17}$$
$$\overset{|}{\underset{}{\text{CONR}_2}}$$

Under biphase conditons (in dry dioxane) the activity of *46* for the reaction of 1-bromoheptane with sodium phenoxide decreased with an increase in the length of aliphatic chains. Under triphase conditions, however, only catalyst *46* with n-octyl groups was active for chloride displacement reactions, and catalysts with groups smaller than n-butyl showed no activity [165]. Polyacrylamides *46* with short aliphatic chains are too hydrophilic to act as catalysts under triphase conditions. Similar behavior has been observed for ammonium salts supported on dextran [95].

Polymer-supported amine oxides *47–49* were catalytically active for the reaction of 1-bromooctane with aqueous sodium cyanide [166].

$$47 : n = 2$$
$$48 : n = 3$$

$$49$$

The catalysts were conditioned in a 1-cyanooctane/aqueous NaCN mixture for 24 h at room temperature to avoid the induction period of the reaction. Rates (converted from a weight basis to a molar basis) with catalysts *47-49* significantly decreased as the % RS increased over the range 5 % to 50 %. With equal loadings activities of the polymer-supported amine oxides decreased with decreased lipophilicity of the catalysts (*49 > 48 > 47*). Lipophilic character appears to be an important factor for activity of polymer-supported cosolvents.

Polymer-supported poly(ethylene glycol) analogues 50 [145, 156, 167, 168] and 51 [167] were effective catalysts for hydroxide, iodide, and phenoxide displacement reactions, but not for cyanide, chloride, and acetate displacements [169]. These catalysts are highly active for various solid/solid/liquid phase transfer reactions (Sect. 6).

50 : R = H
51 : R = CH$_3$

The triphase hydrolysis of 1-bromoadamantane with catalysts 50 (n = 1–16) and 51 was studied kinetically [170]. The enthalpies of activation (ΔH^+) for the catalyzed reactions were 6–12 kcal/mol lower than for the uncatalyzed reaction. (The free energies of activation were 1–2 kcal/mol lower). This considerable variation in ΔH^+ was attributed to a much different microenvironment in catalysts 50 and 51 from that in the absence of the catalyst. Thus Regen [170] pictures both the organic phase and the aqueous phase in the polymer matrix, as if homogeneous. Studies using spin-labeled compounds also indicated that catalyst 50 affects the polarity and the motional freedom of the microenvironment [171].

Recently, catalyst 50 (n > 4) was reported highly active and selective for olefin synthesis from alkyl halides with aqueous sodium or potassium hydroxide without the formation of by-product alcohols [172]. The active catalyst structures were suggested to involve self-solvated polymeric alkoxides [173] 52 and/or complexed hydroxides 53.

52   53

When the reactions of alkyl bromides (n-C$_4$-C$_8$) with phenoxide were carried out in the presence of cosolvent catalyst 51 (n = 1 or 2, 17% RS) under triphase conditions without stirring, rates increased with decreased chain length of the alkyl halide [82]. The substrate selectivity between 1-bromobutane and 1-bromooctane approached 60-fold. Lesser selectivity was observed for polymer-supported HMPA analogue 44 (5-fold), whereas the selectivity was only 1.4-fold for polymer-supported phosphonium ion catalyst 1. This large substrate selectivity was suggested to arise from differences in the effective concentration of the substrates at the active sites. In practice, absorption data showed that polymer-supported poly(ethylene glycol) 51 and HMPA analogues 44 absorbed 1-bromobutane in preference to 1-bromooctane (6–7% excess), while polymer-supported phosphonium ion catalyst 1 absorbed both bromides to nearly the same extent.

To clarify mechanisms of substrate selectivity, studies on elementary reaction steps with polymer-supported cosolvent catalysts must be carried out in detail.

6 Solid/Solid/Liquid Phase Transfer Catalysis

There is a growing of examples of the use of polymer-bound phase transfer catalysts with organic reactants in solution and the inorganic reactant in the form of undissolved solid powder. As with the more common solid/liquid/liquid systems the first examples reported were nucleophilic displacement reactions.

McKenzie and Sherrington [174] found that mixtures of solid sodium or potassium phenoxide, 1-bromobutane, and poly(ethylene glycol) bound to cross-linked polystyrene (54) in refluxing toluene gave good yields of butyl phenyl ether (Eq. (10)).

$$\text{P}\!\!-\!\!\bigcirc\!\!-\!\!CH_2O(CH_2CH_2O)_{30}Ph \qquad 54$$

$$C_6H_5OK(s) + n\text{-}C_4H_9Br \longrightarrow C_6H_5OC_4H_9 + KBr \tag{10}$$

In a typical experiment 0.48 mmol of 1-bromobutane in 2 ml of toluene and catalyst containing 1.16 mmol of ethereal oxygen were stirred magnetically with 2 mmol of powdered sodium phenoxide at reflux for 5 h. In the absence of catalyst there was little or no reaction. With aqueous rather than powdered solid sodium phenoxide the reaction was slow. Soluble as well as polystyrene-bound poly(ethylene glycols) were effective catalysts, as shown in Table 11. It made little difference whether the sodium phenoxide was dried before use or was deposited on the surface of the polymer beads. Potassium phenoxide was slightly more active than sodium phenoxide. Catalyst activity increased as the number of ethylenoxy units in the grafted chain increased from 3 to 9 to 30 (with corresponding decreases in the % RS of the polystyrene) [174]. Onium ion catalysts such as the cellulose-based 55, 7% substituted on saccharide segments, also were active for preparation of butyl phenyl ether [175].

$$\text{C}\!\!-\!\!O\overset{O}{\overset{\|}{C}}CH_2\!\!-\!\!\overset{\oplus}{P}(n\text{-}C_4H_9)_3\,\overset{\ominus}{Cl} \qquad 55: \text{C} = cellulose$$

Yanagida [176] reported similar results for the reaction of 1-bromooctane with alkali iodides in refluxing benzene (Eq. (4)). No reaction occured in the absence of solid catalyst. A mixture of 1.6 mmol of 1-bromooctane, 1 ml of benzene and 5 mol % of

Table 11. Reactions of Phenoxide Ion with 1-Bromobutane in Refluxing Toluene[174]

Catalyst, 2% CL	$M^{+-}OPh$	$10^5\,k_2$, $M^{-1}s^{-1}$
none	solid NaOPh or KOPh	<0.1
none	aq NaOPh	0.26
Ph[OCH$_2$CH$_2$]$_{30}$OH	solid KOPh	16
54, 3% RS	solid KOPh	40
54, 8% RS	solid KOPh	28
54, 8% RS	solid NaOPh	23
54, 3% RS	aq NaOPh	1.2

poly(ethylene glycol) bound to cross-linked polystyrene (51) stirred magnetically at reflux for 1–5 h gave good yields of 1-iodooctane. Catalytic activity increased as the number of ethylenoxy units grafted to the polystyrene increased from 3 to 6 to 8. Powdered sodium iodide gave a reaction rate ten times faster than aqueous sodium iodide. Sodium iodide was more active than potassium iodide. The soluble catalyst, 18-crown-6 ether, was three times more active than the best of the polystyrene-bound poly(ethylene glycols), as shown in Table 12.

Table 12. Reactions of Iodide Ion with 1-Bromooctane in Refluxing Benzene[176]

Catalyst, 24% RS, 2% CL	M^+I^-	$10^5\ k,\ s^{-1}$
none	solid NaI	<0.01
51, n = 3	solid NaI	1.8
51, n = 6	solid NaI	37
51, n = 6 (recycled)	solid NaI	20
51, n = 8	solid NaI	44
51, n = 6	aq NaI	3.1
51, n = 8	solid KI	4.2
51, n = 6	aq KI	2.5
18-crown-6	solid NaI	130

Since mass transfer and intraparticle diffusion limit most rates of reactions in triphase solid/liquid/liquid catalysis, one might expect them to be even more important when a reactant must transfer from one insoluble solid to an active site in another insoluble solid. The cross-linked polymers were vital to the reactions, but the mechanism by which the inorganic reactant, potassium phenoxide in the McKenzie and Sherrington experiments or sodium iodide in the Yanagida experiments, is transferred to the catalyst is unknown. Perhaps the inorganic salts were slightly soluble in the organic solvents and dissolved rapidly from the powder surface, transferred through the liquid phase to the polymer surface, and diffused to the active sites in the polymer gel. Alternatively the inorganic salt might have been transferred from the powder surface to the polymer surface without ever passing through the liquid phase when the two solids came in contact in the stirred mixture. Both Sherrington and Yanagida have carried out experiments designed to answer the question of how the inorganic salt transferred from powder to polymer.

The kinetics of the reaction of solid sodium iodide with 1-bromooctane were studied with a 95% RS graft of poly(ethylene oxide) 6-mer methyl ether on 3% CL polystyrene as catalyst (51)[176]. The rates were approximately first order in 1-bromooctane and independent of the amount of excess sodium iodide. The rates varied with the amount of the solid catalyst used, but there was not enough data to establish the exact functional dependence. All experiments employed powdered sodium iodide, magnetic stirring, and 75–150 μm catalyst beads. Thus the variables stirring speed and particle size, which normally are affected by mass transfer and intraparticle diffusion, were not studied. Yanagida[177] favors a mechanism of transfer of the sodium iodide by dissolution in the solvent (benzene) and diffusion to the catalyst particle surface, because sodium iodide is much more soluble in benzene (0.3 mM) than potassium iodide, accounting for the greater reactivity of sodium iodide. Similarly in the catalysis of Michael addition of nitroethane to acrylonitrile by flouride ion (Eq. (11)) and in the preparation of a phenacyl ester by fluoride ion catalysis (Eq. (12)) no reaction occurred when the highly effective potassium flouride was replaced by sodium fluoride. Sodium fluoride was assumed to be less soluble than potassium fluoride in benezene, although no supporting data was cited.

$$CH_3CH_2NO_2 + CH_2\!\!=\!\!CHCN \xrightarrow{\text{KF}} CH_3C(CH_2CH_2CN)_2\,NO_2 \tag{11}$$

$$C_2H_5CO_2H \;+\; BrCH_2CO\!-\!\!\!\bigcirc\!\!\!-Br \xrightarrow{\text{KF}} C_2H_5CO_2CH_2CO\!-\!\!\!\bigcirc\!\!\!-Br \tag{12}$$

McKenzie and Sherrington [178] have made the most detailed mechanistic study of a solid/solid/liquid reaction. The kinetics of reaction between solid potassium phenoxide and 1-bromobutane (Eq. (10)) catalyzed by 51 (n = 3) were first order in 1-bromobutane, independent of the amount of solid potassium phenoxide and directly proportional to the weight of catalyst used. The dependence on 1-bromobutane concentration and the lack of dependence on the amount of solid potassium phenoxide indicate that dissolution of the solid reactant could not have been rate limiting, and that either diffusion or chemical reaction of the organic reactant was rate limiting. Diffusional limitation was ruled out on the basis of experiments with polystyrenes having different % CL, % RS, particle sizes, and macroporous morphology. The polymers varied in particle size from 15–20 µm to 250–500 µm, in % CL from 2% to 42% divinylbenzene, and in % RS from 29% to 71%, yet the reaction rate constants were all the same. A study of the relative reaction rates of 1-bromobutane and 1-bromooctane with solid potassium phenoxide showed no difference between soluble polymer catalysts and cross-linked polymer catalysts [179]. All of these results point to a rate-limiting chemical reaction between the 1-bromobutane and the phenoxide ion at sites in the various polymers which have nearly equal activity in spite of the wide variations in polymer structures.

Further experiments were carried out to determine the mechanism by which potassium phenoxide is transferred from the solid powder to the catalyst [178]. The solubility of potassium phenoxide in toluene was determined to be <4 µM. Careful washing of the catalysts and recycling of washed catalyst ruled out the presence of a trace of soluble poly(ethylene glycol) which might have transported potassium phenoxide from the solid surface to the polymer surface. In a special partitioned cell reactor a solution of 1-bromobutane with solid potassium phenoxide was separated by a glass frit from an identical solution with the solid catalyst but no potassium phenoxide. With rapid magnetic stirring in both sides of the reactor the rate of reaction of phenoxide with 1-bromobutane in the side of the reactor which had no catalyst was no faster than in control experiments in a standard reactor with no catalyst present. These experiments were interpreted by mechanism in which transfer of potassium phenoxide to the catalyst surface occurred by direct contact between the two insoluble solids.

Trace amounts of water in the 1-bromobutane/potassium phenoxide mixture aided the reaction as shown in Table 13 [178]. Specially dried reactants (by methods

Table 13. Effect of Added Water on the Rate of Reaction of
1-Bromobutane with Potassium Phenoxide[178]

µL (mmol) water added	$10^5\ k$, $M^{-1}s^{-1}$, 90 °C
"normal"	5.0
"anhydrous"	3.7
20(1.1)	5.5
100(5.5)	3.3
2000(110)	0.1

$[\text{n-C}_4\text{H}_9\text{Br}]_{\text{initial}} = 0.24\,\text{M}$ in 2.0 ml of toluene, KOPh
$= 2.0$ mmol, catalyst $= 51$, n $= 3.2\%$ CL, 3% RS.

not reported) led to slower reactions than the normal reagents, and addition of
traces of water (1.1 mmol compared with 2.0 mmol of potassium phenoxide) to the
"anhydrous" mixture increased the rate constant to slightly higher than that determined under "normal" conditions. The exact location of the water in the reaction
mixture was unknown, but the polar environments inside or on the surfaces of the
potassium phenoxide powder and the catalyst were the most likely sites. Water on the
solid surface(s) might have been responsible for rapid transfer of potassium phenoxide
to the catalyst. The importance of the potassium phenoxide surface was revealed in
other control experiments using large $1.0 \times 0.5 \times 0.03$ cm potassium phenoxide
crystals, which gave slower reaction rates under non-stirred conditions than the
powdered potassium phenoxide [178]. The lower surface area of the large crystals made
mass transfer of potassium phenoxide from the solid rate limiting. In refluxing toluene
with no alkyl halide the large crystals of potassium phenoxide dissolved into the
polymeric catalyst up to a concentration of 0.44 mole of salt per mole of polyether
in only 15 min. McKenzie and Sherrington [178] determined carefully the experimental
factors that control one solid/solid/liquid process, the reaction of 1-bromobutane
with solid potassium phenoxide catalyzed by polystyrene-bound polyethers, but they
were properly cautious about extending their conclusions to other systems.

A variety of other substituents on the chain ends of poly(ethylene glycol) grafted
onto polystyrene have been investigated for catalysis of the reaction of 1-bromobutane
and solid potassium phenoxide. The apparent activities of 2% CL, 14% RS catalysts
56 (n = 4) were Z = 2-(1,4-benzodioxanylmethyl) < tetrahydrofuran-2-yl < 2-pyri-
dylmethyl < p-toluenesulfonyl < 8-quinolyl < 2-pyrrolidinon-1-yl < 2-methoxy-
phenyl < 2-naphthyl [179].

$$\text{P}-\text{\textcircled{}}-\text{CH}_2\text{O}(\text{CH}_2\text{CH}_2\text{O})_n\,\text{Z} \qquad 56$$

Substrate selectivity effects were investigated with polystyrene-supported poly-
(ethylene glycol) catalysts 51 (n = 3, 61% RS) and 56 (Z = phenyl, n = 30, 8% RS)
under solid/solid/liquid conditions [180]. The difference in rates of reaction of 1-bromo-
butane and 1-bromooctane with solid potassium phenoxide was a factor of about
3 (51 was more active). Measurements of the distribution of both bromides between

the bulk toluene phase and the swollen catalyst *51* in the presence of water indicated that the catalyst absorbed 1-bromobutane preferentially (a 9 % excess).

The nucleophilic displacement reaction of benzyl chloride with solid potassium acetate in various solvents under solid/solid/liquid conditions was faster with the polymer-bound catalyst *57* than with the soluble analog *58* [181] (See Eq. (13)).

$$C_6H_5CH_2Cl + KO_2CCH_3 \longrightarrow C_6H_5CH_2O_2CCH_3 + KCl \qquad (13)$$

In refluxing acetonitrile the activities of catalysts *57–60* were *60* > *59* > *58* > *57*, while in refluxing dioxane *60* > *57* > *58* = *59*. However, the reactions were very slow. In dioxane only the soluble 18-crown-6 ether (*60*) gave a yield of benzyl acetate of more than 31 % in 2 days. Details of catalyst and potassium acetate particle size and stirring method were not reported [182]. The higher activity of the solid catalyst *57* than of the soluble analog *58* is highly unusual because mass transfer and intra-

Table 14. Phase Transfer Catalysts from Crown Ethers *61* and *62* [183]

Catalyst	Mol %					Relative Activity		
	61	*62*	*63*	*p*-DVB	Styrene	Eq. (13), 25 °C	Eq. (13), 50 °C	Eq. (14), 80 °C
P1	0.8		0.2			2		
P2	0.45		0.1		0.45	1		
P3	0.45			0.1	0.45	2		2
P4	0.4			0.2	0.4	3		
P5		0.46		0.05	0.49		1	2
P6		0.45		0.1	0.45		1	2
P7		0.4		0.2	0.4		2	3
60								1

particle diffusional limitations normally retard the rates of reactions with solid catalysts. There must have been a microenvironmental effect within the cross-linked polymer that raised the intrinsic rate of reaction of acetate ion with benzyl chloride in catalyst *57*.

Catalysts synthesized from crown ether monomers *61* and *62* by copolymerization with styrene and either p-divinylbenzene or p,p′-divinylbiphenyl (*63*) are listed in Table 14 along with their relative activities for solid/solid/liquid reactions of potassium acetate with benzyl chloride (Eq. (13)) and potassium cyanide with 1,4-dichlorobutane (Eq. (14)) in acetonitrile [183].

$$ClCH_2CH_2CH_2CH_2Cl \xrightarrow{\text{KCN}} ClCH_2CH_2CH_2CH_2CN \xrightarrow{\text{KCN}} NCCH_2CH_2CH_2CH_2CN \qquad (14)$$

The catalysts were synthesized by solution polymerization and ground into 0.1–0.2 mm particles. Reaction mixtures were stirred magnetically. Catalysts *P5* and *P6* (5% and 10% CL with p-DVB) had the same activity, higher than that of 20% CL *P7*. Similarly with *63* as cross-linker the 10% CL catalyst *P2* was more active than the 20% CL *P1*. The 10% CL catalysts *P2* and *P3* show that the longer cross-linker *63* gave a more active catalyst, either because of faster intraparticle diffusion of reactants or because of a change in the microenvironment. The reactions had half-lives of 30–40 h. If intraparticle diffusion limits the rates, there is little hope for any of these reactions to be completed the short times required for industrial processes, even at higher temperatures. Comparison of *P3* and *P6* shows that a change in the spacer length of *61* and *62* had no appreciable effect on activity.

Numerous examples of solid/solid/liquid phase transfer catalysis are now known to be useful synthetically but have not been investigated mechanistically. Poly(ethylene glycol) immobilized on alumina and silica gel is active for reaction of solid potassium acetate with 1-bromobutane [184]. Some of the best synthetic results with polymer supports are shown in Table 15. Often use of other solid salts or other catalysts gave poorer yields. It would be valuable to know for the design of future syntheses how these reactions depend on the partial solubility of the inorganic salts in the organic solvents and on the presence of trace amounts of water.

Table 15. Examples of Solid/Solid/Liquid Phase Transfer Catalysis with Solid Inorganic Reagents

Reaction	Catalyst	Ref.
	A	[177]
$PhCH_2OH + CHCl_3 \xrightarrow[56\%]{NaOH} PhCH_2Cl$	A	[177]
$PhCONH_2 + CHCl_3 \xrightarrow[100\%]{NaOH} PhCN$	A	[177]
$C_6H_5CO_2H + (CF_3)_2C{=}CFCF_2CF_3$ $\xrightarrow[CH_2Cl_2]{K_2CO_3} C_6H_5COF + R_fO_2CC_6H_5$ 47% 21%	B	[185]
cycloheptanol $\xrightarrow[\substack{CH_2Cl_2 \\ 92\%}]{Ca(OCl)_2}$ cycloheptanone	B	[186]
	C	[187]
		[187]
		[187]
	D	[188]
$n\text{-}C_6H_{11}COCH_3 + NaBH_4 \xrightarrow[\substack{trace\ of\ H_2O \\ 98\%}]{ClCH_2CH_2Cl,} C_6H_{11}CHOHCH_3$	E	[175]

Catalysts: A, 51, 3% CL, 95% RS;
       B, 2, Amberlite IRA-900 [108], macroporous, low % CL;
       C, 2, Amberlite IRA-904 [108], macroporous, high % CL;
       D, Ⓟ—C_6H_4—$CH_2N(CH_3)_2$ (n-C_8H_{17}), Cl^-, 2% CL, 10% RS;
       E, cellulose-$OCOCH_2CH_2\overset{+}{P}$(n-$C_4H_9$)$_3$, Cl^-, 7% substituted

7 Catalyst Stability

The utility of polymer-supported phase transfer catalysts depends upon their ease of synthesis and their chemical and physical stability. The advantages of the heterogeneous catalysts are the ease of separation of the catalyst from reaction mixtures and reuse. Although there may occasionally be cases of higher activity of heterogeneous catalysts than of analogous homogeneous catalysts, in general one should expect lower activity because of diffusional limitations. In alkaline media quaternary ammonium and phosphonium ions decompose by nucleophilic attack at alpha-carbon atoms, by E2 eliminations, and by attack of hydroxide and other oxy anions at positively charged phosphorus. Benzylic quaternary ammonium and phosphonium ions are particularly unstable and cannot be recycled often except for reactions in neutral or acidic media [154]. Inclusion of a spacer chain of at least three carbon atoms between the charged site and the aromatic ring of a polystyrene support greatly enchances the stability, enabling recycling from media as alkaline as concentrated aqueous sodium cyanide [79]. Of cource, the spacer chain catalysts cost more to prepare than do the benzyl onium ion catalysts. If high temperatures are required and basic oxy anions are absent, phosphonium ions are more stable than ammonium ions [116-118].

Crown ether, cryptand, and poly(ethylene glycol) catalysts are more stable in base than the quaternary ammonium and phosphonium ions. Only the poly(ethylene glycols) are likely to meet industrial requirements for low cost, although a number of more efficient, lower cost crown ether syntheses have appeared recently, such as those of sila-crowns *64* bound to silica [189].

64

For general use the ethers of oligo(ethylene glycols) are the most promising polystyrene-supported catalysts. They are more hydrophilic than the most active quaternary onium ions, however, and the resulting lower activity for nucleophilic displacements in solid/liquid/liquid phase transfer systems may require that they be used at higher temperature. A likely drawback to the use of the poly(ethylene glycol) catalysts is air oxidation. Catalysts *51* oxidize at the benzylic carbon atom in a drying oven at 80 °C to produce carboxylic acids as in Eq. (15) with weak acid ion exchange capacities from 0.1 to 2.5 mequiv/g [190].

$$ \text{(P)}\!-\!\!\bigcirc\!\!-\!CH_2O(CH_2CH_2O)_nCH_3 \xrightarrow[\text{[O]}]{\Delta} \text{(P)}\!-\!\!\bigcirc\!\!-\!CO_2H \qquad (15) $$

For long term use physical stability of the catalyst is essential. Cross-linked polystyrenes have been used for years in the form of ion exchange resins and do posses the stability required for most applications. Many reservations about polystyrene stability have been expressed, but we believe they can be overcome. The problem of bead breakage in magnetically stirred mixtures can be avoided by mechanical stirring or shaking. Another problem is the weakness of solvent-swollen polystyrene gels with low degrees of cross-linking. Most academic research has been carried out with 1 % and 2 % CL Merrifield resins, which may swell to five or more times their dry volume under catalysis conditions. Swelling can easily be reduced by use of more highly cross-linked supports, although activity will be lower as long as there are intraparticle diffusional limitations on observed reaction rates. Since the major reason for choosing a heterogeneous catalyst is usually ease of separation or recycling, the lower activity of catalysts on more highly cross-linked supports is often an acceptable compromise. Silica gel and alumina are more stable physically than most synthetic polymers, and they do not swell. Alumina is a particularly promising support for phase transfer catalysts because of its tolerance of basic media, but more research on alumina-supported catalysts is needed to establish their strengths and their limitations.

8 Conclusions

All of the available evidence indicates that triphase reactions with organic liquid, aqueous salt solution, and solid polymeric catalyst are complicated examples of ion exchange resin catalysis. As with other heterogeneous catalysts the observed reaction rates usually are lower than those attained with analogous soluble catalysts because of mass transfer and intraparticle diffusional limitations. The triphase systems are complicated by the requirement for mass transfer from both organic and aqueous phases and intraparticle diffusion of both ions and nonpolar organic reactants. In the few examples that have been studied in detail the mass transfer limitations have been overcome with vigorous stirring. The intraparticle diffusional limitations can be minimized by use of a) as lightly cross-linked, swellable polymer supports as possible within requirements for particle durability, b) as small particles as can be filtered practically, and c) spacer chains to put the active sites in a more fluid environment within the polymer network.

The active site structure must provide high activity and promote transport of ions and nonpolar organic molecules through the polymer network. For most nucleophilic displacement reactions quaternary ammonium or phosphonium ions with chains of C_4 or greater on the cationic center and <25 % RS provide a suitably lipophilic environment for organic transport and for high intrinsic reactivity. Activity is enhanced by the separation of the active site from the polymer backbone by at least a short spacer chain. However, the lipophilic environment and low density of the ionic sites can lead to slow ion exchange. Crown ethers, cryptands, and oligomeric poly(ethylene glycol) convert a formally uncharged polymer into an ion exchange resin under phase transfer catalysis conditions. Cation complexation must be strong for high catalytic activity.

We now have a basic understanding of the heterogeneous catalytic nature of solid/liquid/liquid phase transfer catalysis that will help in the design of large scale processes and in the execution of laboratory scale syntheses. Better understanding of the mass transfer and diffusional processes in solid/solid/liquid phase transfer reactions is needed. In particular, investigation of the mechanisms by which the inorganic salt is transported from the surface of a crystalline solid to the active sites within an insoluble polymeric catalyst would be valuable. Large scale applications of heterogeneous phase transfer catalysts are likely to require continuous flow reactors. At present almost nothing is known about mass transfer in mixtures of aqueous and organic liquids passing through a catalyst bed. Another area worthy of research is the synthesis and use of polymer supports based on more polar monomers than styrene or with more polar grafts on polystyrene to modify the hydrophile/lipophile balance.

Progress is expected in the synthetic uses of solid/solid/liquid systems and in the use of chiral catalysts for asymmetric synthesis.

Polystyrene has been used most often as the support for phase transfer catalysts mainly because of the availability of Merrifield resins and quaternary ammonium ion exchange resins. Although other polymers have attrative features, most future applications of polymer-supported phase transfer catalysts will use polystyrene for several reasons: It is readily available, inexpensive, easy to functionalize, chemically inert in all but strongly acidic media, and physically stable enough for most uses. Silica gel and alumina offer most of these same advantages. We expect that large scale applications of triphase catalysis will use polystyrene, silica gel, or alumina.

9 References

1. Kunitake, T., Shinkai, S.: Adv. Phys. Org. Chem. *17*, 435 (1980)
2. Royer, G. P.: Adv. Catal. *29*, 197 (1980)
3. Overberger, C. G., Mitra, S.: Pure Appl. Chem. *51*, 1391 (1979)
4. Ise, N., Okubo, T., Kunugi, S.: Acc. Chem. Res. *15*, 171 (1982)
5. Gates, B. C., Katzer, J. R., Schuit, G. C. A.: Chemistry of Catalytic Processes. New York: McGraw-Hill 1979
6. Satterfield, C. N.: Heterogeneous Catalysis in Practice. New York: McGraw-Hill 1980
7. Helfferich, F.: Ion Exchange, ch. 11. New York: McGraw-Hill 1962.
8. Pitochelli, A. R.: Ion Exchange Catalysis and Matrix Effects. Philadelphia, Rohm and Haas Co. 1975
9. Lieto, J., et al.: Chemtech. *13*, 46 (1983)
10. Klein, J.: Makromol. Chem., Suppl. *5*, 155 (1981)
11. Heinz, W. E., MacLean, A. F.: U.S.A. Patent 3,007,500 (1963)
12. Cossu-Jouve, M., Savon, M.-C., Ucciani, E.: Bull. Soc. Chim. Fr., 2429 (1973).
13. Astle, M. J., Zaslowsky, J. A.: Ind. Eng. Chem. *44*, 2867 (1952)
14. Gaset, A., Gorrichon, J. P.: Syn. Commun. *12*, 71 (1982)
15. Union Carbide Corp.: U. K. Patent 1,499,137 (1978)
16. Chiellini, E., Solaro, R., D'Antone, S.: Makromol. Chem., Suppl. *5*, 82 (1981)
17. Mastagli, P., Floc'h, A., Durr, G.: Compt. Rend. Acad. Sci., 1402 (1952)
18. Hoffmann, W., Pasedach, H.: German Patent 1,223,364 (1966)
19. Shimo, K., Wakamatsu, S.: J. Org. Chem. *28*, 504 (1963)
20. Schmidle, C. J., Mansfield, R. C.: U.S.A. Patent 2,658,070 (1953)
21. Howk, B. W., Langkammerer, C. M.: U.S.A. Patent 2,579,580 (1951)
22. Copelin, H. B., Crane, G. B.: U.S.A. Patent 2,779,781 (1957)
23. Jeltsch, A. E.: U.S.A. Patent 2,852,566 (1958)
24. Boeva, R., Markov, K., Kotov, St.: J. Catal. *62*, 231 (1980)
25. Pflugfelder, B., Vannel, P.: U.S.A. Patent 3,290,383 (1966)
26. Wheeler, E. N., Stearns, D. L.: U.S.A. Patent 3,340,295 (1967)
27. Mulder, R. J.: Chemisch Weekblad *62*, 421 (1966)
28. Dawydoff, W.: Z. Polymerforsch. *27*, 189 (1976).
29. Linarte Lazcano, R., Germain, J.-E.: Bull. Soc. Chim. Fr., 1869 (1971)
30. Ichikawa, M.: Japan Kokai 76 95,992 (1976)
31. Haag, W. O., Whitehurst, D.: U.S.A. Patent 4,111,856 (1978)
32. Rericha, R., et al.: Coll. Czech. Chem. Comm. *44*, 3183 (1979)
33. Hartwell, G. E., Garrou, P. E.: U.S.A. Patent 4,144,191 (1979)
34. Garrou, P. E., Hartwell, G. E.: U.S.A. Patent 4,262,147 (1981)
35. Waller, F. J.: American Chemical Society Division of Petroleum Chemistry Preprints *27*, 611 (1982)
36. Merrifield, R. B.: J. Am. Chem. Soc. *85*, 2149 (1963)
37. Erickson, B. W., Merrifield, R. B.: In: The Proteins, 3rd ed. (Neurath, H., Hill, R. L., Boeder, C.-L., eds.) vol. II, p. 255. New York: Academic Press 1976

38. Letsinger, R., Kornet, M. J.: J. Am. Chem. Soc. *85*, 3045 (1963)
39. Akelah, A., Sherrington, D. C.: Chem. Rev. *81*, 557 (1981)
40. Frechet, J. M. J.: Tetrahedron *37*, 663 (1981)
41. Akelah, A.: Synthesis, 413 (1981)
42. Hodge, P., Sherrington, D. C.: Polymer-supported Reactions in Organic Synthesis. London: John Wiley and Sons 1980
43. Mathur, N. K., Narang, C. K., Williams, R. E.: Polymers As Aids in Organic Chemistry, London: Academic Press 1980
44. Kraus, M. A., Patchornik, A.: Macromol. Rev. *15*, 55 (1980)
45. Chauvin, Y., Commereuc, D., Dawans, F.: Prog. Polym. Sci. *5*, 95 (1977)
46. Whitehurst, D. D.: Chemtech *10*, 44 (1980)
47. Gates, B. C., Lieto, J.: Chemtech *10*, 195, 248 (1980)
48. Ciardelli, F., et al.: J. Mol. Catal. *14*, 1 (1982)
49. Kaneko, M., Tsuchida, E.: Macromol. Rev. *16*, 397 (1981)
50. Osada, Y., Chiba, T.: Makromol. Chem. *180*, 1617 (1979)
51. Yamazaki, N., et al.: Polym. J. *12*, 231 (1980)
52. Manecke, G., Storck, W.: Angew. Chem., Int. Ed. Engl. *17*, 657 (1978)
53. Regen, S. L.: J. Am. Chem. Soc. *97*, 5956 (1975)
54. Brown, J. M., Jenkins, J.: J. Chem. Soc. Chem. Commun. 458 (1976)
55. Cinquini, M., Colonna, S., Molinari, H., Montanari, F., Tundo, P.: J. Chem. Soc. Chem. Commun. 394 (1976)
56. Regen, S. L.: Angew. Chem., Int. Ed. Engl. *18*, 421 (1979)
57. Weber, W. P., Gokel, G. W.: Phase-Transfer Catalysis in Organic Synthesis. Berlin: Springer 1977
58. Starks, C. M., Liotta, C.: Phase Transfer Catalysis. New York: Academic Press 1978
59. Dehmlow, E. V., Dehmlow, S. S.: Phase Transfer Catalysis. Weinheim: Verlag Chemie 1980
60. Montanari, F., Landini, D., Rolla, F.: Top. Curr. Chem. *101*, 147 (1982)
61. Satterfield, C. N.: Mass Transfer in Heterogeneous Catalysis. Cambridge, Massachusetts: M. I. T. Press 1970
62. Thomas, J. M., Thomas, W. J.: Introduction to the Principles of Heterogeneous Catalysis, ch. 4. New York: Academic Press 1967
63. Ref. 7, ch. 6
64. Helfferich, F.: In: Ion Exchange, vol. 1, p. 65 (Marinsky, J. A., ed.). New York: Marcel Dekker 1966
65. Smith, N. L., Amundson, N. R.: Ind. Eng. Chem. *43*, 2156 (1951)
66. Korus, R., O'Driscoll, K. F.: Can. J. Chem. Eng. *52*, 775 (1974)
67. Veith, W. R., et al.: Chem. Eng. Sci. *28*, 1013 (1973)
68. Kunin, R., Meitzner, E., Bortnick, N.: J. Am. Chem. Soc. *84*, 305 (1962)
69. Millar, J. T., Smith, D. G., Marr, W. E., Kressman, T. R. E.: J. Chem. Soc. 218 (1963)
70. Seidl, J., Malinsky, J., Dusek, K., Heitz, W.: Adv. Polym. Sci. *5*, 113 (1967)
71. Guyot, A., Bartholin, M.: Prog. Polym. Sci. *8*, 277 (1982)
72. Frisch, N. W., Chem. Eng. Sci. *17*, 735 (1962)
73. Tomoi, M., Ford, W. T.: J. Am. Chem. Soc. *103*, 3821 (1981)
74. Molinari, H., Montanari, F., Quici, S., Tundo, P.: J. Am. Chem. Soc. *101*, 3920 (1979)
75. Montanari, F.: pers. comm.
76. Regen, S. L., Besse, J. J.: J. Am. Chem. Soc., *101*, 4059 (1979)
77. Regen, S. L., Bolikal, D., Barcelon, C.: J. Org. Chem. *46*, 2511 (1981)
78. Ohtani, N., Wilkie, C. A., Nigam, A., Regen, S. L.: Macromoleculares *14*, 516 (1981)
79. Tomoi, M., et al.: J. Polym. Sci., Polym. Chem. Ed. *20*, 3015 (1982)
80. Tomoi, M., et al.: J. Polym. Sci., Polym. Chem. Ed. *20*, 3421 (1982)
81. Serita, H., Ohtani, N., Kimura, C.: Kobunshi Ronbunshu *35*, 203 (1978)
82. Regen, S. L., Nigam, A.: J. Am. Chem. Soc. *100*, 7773 (1978)
83. Ford, W. T.: J. Polym. Sci., Polym. Chem. Ed. in press.
84. Tomoi, M., Ford, W. T.: J. Am. Chem. Soc. *103*, 3828 (1981)
85. Tomoi, M., Ford, W. T.: J. Am. Chem. Soc. *102*, 7140 (1980)
86. Ford, W. T., Lee, J., Tomoi, M.: Macromolecules *15*, 1246 (1982)

87. Serita, H., et al.: Kobunshi Ronbunshu *36*, 527 (1979)
88. Tomoi, M., Hosokawa, Y., Kakiuchi, H.: Makromol Chem. Rapid Commun. *4*, 22X (1983)
89. Regen, S. L.: J. Am. Chem. Soc. *98*, 6270 (1976)
90. Chiles, M. S., Reeves, P. C.: Tetrahedron Lett. 3367 (1979)
91. Chiles, M. S., Jackson, D. D., Reeves, P. C.: J. Org. Chem. *45*, 2915 (1980)
92. Ohtani, N., Regen, S. L.: Macromolecules *14*, 1594 (1981)
93. Molinari, H., Montanari, F., Tundo, P.: J. Chem. Soc., Chem. Commun. 639 (1977)
94. Saegusa, T., et al.: Polym. J. *11*, 1 (1979)
95. Kise, H., Araki, K., Seno, M.: Tetrahedron Lett. *22*, 1017 (1981)
96. Makosza, M., Bialecka, E.: Tetrahedron Lett. 183 (1977)
97. Dehmlow, E. V., Slopianka, M., Heider, J.: Tetrahedron Lett. 2361 (1977)
98. Solaro, R., D'Antone, S., Chiellini, E.: J. Org. Chem. *45*, 4179 (1980)
99. Kimura, C., Kashiwaya, K., Murai, K.: Asahi Garasu Kogyo Gijutsu Shoreikai Kenkyu Hokoku *26*, 163 (1975)
100. Ref. 58, pp. 64–65 and references therein
101. Komeili-Zadeh, H., Dou, H. J.-M., Metzger, J.: J. Org. Chem. *43*, 156 (1978)
102. Chiellini, E., Solaro, R., D'Antone, S.: Makromol. Chem. *178*, 3165 (1977)
103. Balakrishnan, T., Ford, W. T.: J. Org. Chem. *48*, 1029 (1983)
104. Balakrishnan, T., Ford, W. T.: Tetrahedron Lett. 4377 (1981)
105. Dowex: Ion Exchange. Midland, Michigan: The Dow Chemical Co. 1964
106. Brown, J. M.: In: Further Perspectives in Organic Chemistry, Ciba Foundation Symposium *53*, p. 149. Amsterdam: Elsevier 1978
107. Montanari, F., Quici, S., Tundo, P.: J. Org. Chem. *48*, 199 (1983)
108. Kun, K. A., Kunin, R.: J. Polym. Sci. C *16*, 1457 (1967)
109. Kolarz, B. N., Wieczorek, P. P., Wojaczynska, M.: Angew. Makromol. Chem. *96*, 193 (1981)
110. Amberlite XAD Macroreticular Adsorbents. Philadelphia: The Rohm and Haas Co. 1972
111. Davankov, V. A., Rogozhin, S. V., Tsyurupa, M. P.: In: Ion Exchange and Solvent Extraction, vol. 7, p. 29. Marinsky, J. A., Marcus, Y.: Eds. New York: Marcel Dekker 1977
112. Tundo, P.: Synthesis 315 (1978)
113. Noguchi, H., Sugawara, M., Uchida, Y.: Polymer *21*, 861 (1980)
114. Tundo, P.: J. Chem. Soc., Chem. Commun., 641 (1977)
115. Rolla, F., Roth, W., Horner, L.: Naturwiss. *64*, 337 (1977)
116. Tundo, P., Venturello, P.: J. Am. Chem. Soc. *101*, 6606 (1979)
117. Tundo, P., Venturello, P.: J. Am. Chem. Soc. *103*, 856 (1981)
118. Tundo, P., Venturello, P., Angeletti, E.: J. Am. Chem. Soc. *104*, 6551 (1982)
119. Tundo, P., Venturello, P., Angeletti, E.: J. Am. Chem. Soc. *104*, 6547 (1982)
120. Ragaini, V., Saed, G.: Z. Phys. Chem. *119*, 117 (1980)
121. Tundo, P.: J. Org. Chem. *44*, 2048 (1979)
122. Angeletti, E., Tundo, P., Venturello, P.: J. Chem. Soc., Chem. Commun. 1127 (1980)
123. Angeletti, E., Tundo, P., Venturello, P.: J. Chem. Soc., Perkin Trans I, 993 (1982)
124. Regen, S. L.: J. Org. Chem. *42*, 875 (1977)
125. Julia, S., Masana, J., Vega, J. C.: Angew. Chem., Int. Ed. Engl. *19*, 929 (1980)
126. Ohashi, Inoue, S.: Makromol. Chem. *150*, 105 (1971)
127. Ohashi, S., Inoue, S.: Makromol. Chem. *160*, 69 (1972)
128. Colonna, S., Fornasier, R., Pfeiffer, U.: J. Chem. Soc., Perkin Trans. I, 8 (1978)
129. Yamashita, T., et al.: Bull. Chem. Soc. Japan *51*, 1183 (1978)
130. Yamashita, T., Yasueda, H., Nakamura, N.: Bull. Chem. Soc. Japan *51*, 1247 (1978)
131. Kobayashi, N., Iwai, K.: J. Am. Chem. Soc. *100*, 7071 (1978)
132. Kobayashi, N., Iwai, K.: J. Polym. Sci., Polym. Chem. Ed. *18*, 923 (1980)
133. Kobayashi, N., Iwai, K.: Macromolecules *13*, 31 (1980)
134. Kobayashi, N., Iwai, K.: Tetrahedron Lett. 2167 (1980)
135. Kobayashi, N., Iwai, K.: Makromol. Chem., Rapid Comm. *2* 105 (1981)
136. Sherrington, D. C., Solaro, R., Chiellini, E.: J. Chem. Soc., Chem. Commun. 1103 (1982)
137. Sherrington, D. C., Kelly, J.: Polymer Preprints *23*(1), 177 (1982)
138. Chiellini, E., D'Antone, S., Solaro, R.: Polymer Preprints *23*(1), 179 (1982)
139. Cainelli, G., Manescalchi, F., Contento, M., in: Organic Synthesis Today and Tomorrow (Eds.) Trost, B. M., Hutchinson, C. R.: p 19. New York: Pergamon Press 1981

140. Tomoi, M., et al.: Tetrahedron Lett. 3030 (1978)
141. Tomoi, M., Kihara, K., Kakiuchi, H.: Tetrahedron Lett. 3485 (1979)
142. Montanari, F., Tundo, P.: Tetrahedron Lett. 5055 (1979)
143. Bogatskii, A. V., Lukyanenko, N. G., Pastushok, V. N.: Dokl. Akad. Nauk. SSSR 247, 1153 (1979)
144. Blasius, E., et al.: J. Chromatogr. 201, 147 (1980)
145. Fukunishi, K., Bronislaw, C., Regen, S. L.: J. Org. Chem. 46, 1218 (1981)
146. Montanari, F., Tundo, P.: J. Org. Chem. 46, 2125 (1981)
147. Montanari, F., Tundo, P.: J. Org. Chem. 47, 1298 (1982)
148. Blasius, E., et al.: Makromol. Chem. 183, 1401 (1982)
149. Tomoi, M., et al.: unpubl. results
150. Landini, D., Maia, A., Montanari, F.: J. Chem. Soc., Perkin Trans. II 46 (1980)
151. Landini, D., et al.: J. Chem. Soc., Perkin Trans. II 821 (1981)
152. Pedersen, C. J., Frensdorff, H. K.: Angew. Chem., Int. Ed. Engl. 11, 16 (1972)
153. Lehn, J.-M.: Accts. Chem. Res. 11, 49 (1978)
154. Dou, J. M., et al.: J. Org. Chem. 42, 4275 (1977)
155. Tomoi, M., et al.: Chem. Lett. 473 (1976)
156. Regen, S. L., Dulak, L.: J. Am. Chem. Soc. 99, 623 (1977)
157. Normant, H.: Angew. Chem. 79, 1029 (1967); Int. Ed. Engl. 6, 1046 (1967)
158. Martin, D., Weise, A., Niclas, H.-J.: Angew. Chem. 79, 340 (1967); Int. Ed. Engl. 6, 318 (1967)
159. Vaughn, W.: In: Amides, (Lagowski, J. J., ed.) Vol. II, p. 192. New York: Academic Press 1967
160. Sowinski, A. F., Whitesides, G. M.: J. Org. Chem. 44, 2369 (1979)
161. Parker, A. J.: Chem. Rev. 69, 1 (1969)
162. Regen, S. L., Nigam, A., Besse, J. J.: Tetrahedron Lett. 2757 (1978)
163. Tomoi, M., Ikeda, M., Kakiuchi, H.: Tetrahedron Lett. 3757 (1978)
164. Tomoi, M. et al.: Bull. Chem. Soc. Japan 52, 1653 (1979)
165. Regen, S. L., Mehrotra, A., Singh, A.: J. Org. Chem. 46, 2182 (1981)
166. Maeda, H., Hayashi, Y., Teramura, K.: Chem. Lett. 677 (1980)
167. McKenzie, W. M., Sherrington, D. C.: J. Chem. Soc., Chem. Commun. 541 (1978)
168. Yanagida, S., Takahashi, K., Okahara, M.: Yukagaku 28, 14 (1979)
169. Regen, S. L., Heh, J. C. K., McLick, J.: J. Org. Chem. 44, 1961 (1979)
170. Regen, S. L., Besse, J. J., McLick, J.: J. Am. Chem. Soc. 101, 116 (1979)
171. Regen, S. L.: J. Am. Chem. Soc. 99, 3838 (1977)
172. Kimura, Y., Regen, S. L.: J. Org. Chem. 48, 195 (1983)
173. Bartsch, R. A., Roberts, D. K.: Tetrahedron Lett. 321 (1977)
174. McKenzie, W. M., Sherrington, D. C.: J. Chem. Soc., Chem. Commun. 541 (1978)
175. Akelah, A., Sherrington, D. C.: Eur. Polym. J. 18, 301 (1982)
176. Yanagida, S., Takahashi, K., Okahara, M.: Yukagaku 28. 14 (1979)
177. Yanagida, S., Takahasi, K., Okahara, M.: J. Org. Chem. 44, 1099 (1979)
178. MacKenzie, W. M., Sherrington, D. C.: Polymer 21, 791 (1980)
179. Heffernan, J. G., MacKenzie, W. M., Sherrington, D. C.: J. Chem. Soc., Perkin Trans. II, 514 (1981)
180. MacKenzie, W. M., Sherrington, D. C.: Polymer 22, 431 (1981)
181. Hiratani, K., Reuter, P., Manecke, G.: J. Mol. Catal. 5, 241 (1979)
182. Hiratani, K., Reuter, P., Manecke, G.: Isr. J. Chem. 18, 208 (1979)
183. Manecke, G., Kramer, A.: Makromol. Chem. 182, 3017 (1981)
184. Sawicki, R. A.: Tetrahedron Lett. 23, 2249 (1982)
185. Yanagida, S., Noji, Y., Okahara, M.: Bull. Chem. Soc. Japan 54, 1151 (1981)
186. Schneider, M., Weber, J.-V., Faller, P.: J. Org. Chem. 47, 364 (1982)
187. Hodge, P.: Lecture Internat. Symp. Polymer Supported Reagents in Organic Chemistry, Lyon, France, June 29 — July 1, 1982, and pers. comm.
188. Moore, G. G., Foglia, T. A., McGahan, T. J.: J. Org. Chem. 44, 2425 (1979)
189. Arkles, B., Peterson, W. R., Jr., King, K.: ACS Symp. Ser. 192, 281 (1982)
190. Yanagida, S., Fujita, H., Okahara, M.: Chem. Lett. 1617 (1981)

Received March 22, 1983
T. Saegusa (Editor)

Rapid Polymer Transport in Concentrated Solutions

W. D. Comper and B. N. Preston
Biochemistry Department, Monash University, Clayton, Victoria,
Australia, 3168

This article discusses the rapid transport of polymers in terms of diffusional or diffusional-related processes. In particular it reviews:
1. Polymer and solvent diffusion in binary systems.
2. Polymer transport in ternary systems including an analysis of the cross diffusion coefficients and component distribution within the systems.
3. Macromolecular transport in transient polymeric networks and an introduction to the observation of rapid migration in such systems.
4. Detailed discussion of measurement of and factors involved in rapid polymer transport in multicomponent systems.
5. Identification of structured flows associated with rapid polymer transport and some simple mechanistic interpretations.
6. Rapid polymer transport and model biological systems.
The aim of the article is to introduce new observations of diffusive-convective phenomena in polymer chemistry. The processes discussed are of significance to those interested in transport phenomena.

List of Symbols

A	constant
A	surface area
A_n	nth virial coefficient
A_t	absorbance at 237 nm at time t
C	concentration in mass/volume units
C_A, C_B	concentration above and below the boundary, respectively
C°	initial concentration
C^*	critical polymer concentration
$(D)^\circ$	diffusion coefficient at infinite dilution
D_{coop}	co-operative diffusion coefficient
D_i^+	intradiffusion or self-diffusion coefficient
$(D)_v$	mutual diffusion coefficient in a volume-fixed frame of reference.
D_{ii}, D_{ij}	principal and cross-diffusion coefficients, respectively
$Đ_{ij}$	Stefan-Maxwell diffusion coefficient
f_{ij}	frictional coefficient per mole of i
J_i	flux of i
$(J_i)_v$	flux of i in a volume-fixed reference frame
L	concentration of rods expressed as cm polymer per cm^3 solution
L_{ij}	phenomemological coefficients
m	molar concentration
M	molecular weight (number average)
$\bar{M}_n$	number average molecular weight
$\bar{M}_w$	weight average molecular weight
Q	amount transported over a boundary
r	radius of diffusing particle
R	universal gas constant
S	sedimentation coefficient
t	time
T	temperature
T	transport coefficient
T^c	transport coefficient obtained by diffusional analysis employing the open-ended capillary technique
v	velocity
V_c	critical volume
V_i	molar volume of i
$\bar{V}_i$	partial specific volume of i
x_i	fraction of species i
ζ	frictional coefficient
η	viscosity
η_{rel}	relative viscosity
η_0	solvent viscosity
μ_i	chemical potential of i
ξ	distance between successive contact points in a transient network or 'length of blob'

Π   osmotic pressure
ϱ   density

Abbreviations

PEG   poly(ethyleneglycol)
PVA   poly(vinyl alcohol)
PVP   poly(vinylpyrrolidone)
dextran Tn   dextran, $\bar{M}_w \simeq 1000 \times n$
HTO   Tritiated water

1 Introduction

Whereas polymer transport in dilute solution has been widely studied, investigations of polymer transport in concentrated solutions have been reported only infrequently. When solutions of high molecular weight polymers are concentrated in a way that intermolecular interactions between individual molecules occur, it becomes increasingly evident that both the static and dynamic properties of the macromolecules may be markedly altered compared to their behaviour in dilute solution [1].

Under certain circumstances, polymer transport may procced rapidly in concentrated solutions. Our stimulus for investigating the transport properties in concentrated solutions comes from attempts to understand the transport phenomena in biological systems which can be generally characterised as concentrated polymer systems [1,2]. For example, the space between cells in tissues is occupied by an insoluble fibre network of collagen which has interdispersed within it a soluble phase containing high concentrations of high molecular weight polysaccharide. Therefore, molecular transport in tissues, to and from cells, is influenced and perhaps governed by the presence of this soluble concentrated polysaccharide solution.

In view of the intrinsic complexity and variability of biological systems, we have selected simple model systems containing polymers in order to gain optimal mechanistic interpretations. Progress in polymer transport studies in these model systems of concentrated polymer solutions has encompassed both binary and ternary systems. These models are not only of tissue systems but appear to have general applicability. It is hoped that our recent investigations and findings discussed in this applicability. It is hoped that our recent investigations and findings discussed in this article, will find application in industry or technology. It is clear that by developing and understanding the basic parameters associated with the dynamic properties of multicomponent polymer-containing systems, we will eventually delineate those areas in which such information is commercially valuable.

This paper will deal primarily with rapid transport derived from diffusion processes in aqueous solution. These processes may be observed in simple polymer, water systems following well-established thermodynamic principles. In particular, we shall discuss ternary polymer-containing systems in which very rapid transport processes, associated with the formation of macroscopic structures in solution, occur.

2 Polymer Transport in Binary Systems

2.1 Basic Theoretical Considerations

The simplest system used for diffusional analysis is that of an isothermal, isobaric binary system where the micromolecular solvent (H_2O) is designated as component 1 and the solute as component 2. Thus, for concentration gradients of these components, we may measure the net flux of solute across an arbitrary plane or boundary due to the relaxation of the concentration gradient. The interdiffusional flux in a binary liquid mixture is commonly described as mutual diffusion.

Diffusion equations which are normally used for the experimental analysis of the relaxation of the concentration gradient (i.e. practical diffusion equations) describe the fluxes J_1 and J_2 of components 1 and 2 respectively as

$$(J_1)_v = - (D_{11})_v \, \text{grad} \, m_1 \tag{1}$$

$$(J_2)_v = - (D_{22})_v \, \text{grad} \, m_2 \tag{2}$$

where the letter v denotes all quantities relative to a volume-fixed frame of reference. $(D_{ii})_v$ is the diffusion coefficient and m_i the molar concentration of component i. The volume-fixed reference frame requires that there are zero-volume flows, that is

$$\Sigma \, (J_i)_v V_i = 0 \tag{3}$$

where V_i is the molar volume of i. The Gibbs-Duhem equation places a restriction on the isothermal concentration gradients in the mechanical equilibrium such that

$$V_1 \, \text{grad} \, m_1 + V_2 \, \text{grad} \, m_2 = 0 \tag{4}$$

With Eqs. (1)–(4) it is easy to show that

$$(D_{22})_v = (D_{11})_v = (D)_v \tag{5}$$

where $(D)_v$ is the single mutual diffusion coefficient which suffices to describe the interdiffusion of both components (this diffusion coefficient is sometimes referred to as the osmotic diffusion coefficient).

The phenomenological treatment allows us to identify the diffusion coefficients with the gradients in chemical potentials such that [3]

$$(D_{11})_v = (L_{11})_v \, \partial\mu_1/\partial m_1 \tag{6}$$

$$(D_{22})_v = (L_{22})_v \, \partial\mu_2/\partial m_2 \tag{7}$$

where L_{11} and L_{22} are the phenomenological coefficients and μ_i is the chemical potential of component i. For further development of these equations it is useful to introduce the frictional coefficient formalism of Spiegler [4] as presented by Kedem and Katchalsky [5] which takes the form

$$-\frac{\partial \mu_1}{\partial x} = f_{12}(v_1 - v_2) \tag{8}$$

$$-\frac{\partial \mu_2'}{\partial x} = f_{21}(v_2 - v_1) \tag{9}$$

where v_i is the velocity of component i and f_{ij} the frictional coefficient per mole of i. The velocity difference, $(v_1 - v_2)$, is independent of the choice of the reference frame, so that the same independence applies to frictional coefficients f_{ij} (see Footnote 1). The frictional coefficients follow the reciprocal relationship

$$m_1 f_{12} = m_2 f_{21} \tag{10}$$

Knowing that $J_i = m_i v_i$ and using Eq. (3), it can be shown from Eqs. (8) and (9) that

$$(D_{11})_v = \frac{m_1}{f_{12}\left(1 + \dfrac{m_1 V_1}{m_2 V_2}\right)} \frac{\partial \mu_1}{\partial m_1} \tag{11}$$

and

$$(D_{22})_v = \frac{m_2}{f_{21}\left(1 + \dfrac{m_2 V_2}{m_1 V_1}\right)} \frac{\partial \mu_2}{\partial m_2} \tag{12}$$

---

1 Another frictional coefficient formalism is that represented by the generalised Stefan-Maxwell equation [6]

$$\frac{\partial \mu_i}{\partial x} = RT \, \Sigma \, \frac{x_i}{Ð_{ij}} (v_j - v_i)$$

where x_j is the fraction of species j and $Ð_{ij}$ the Stefan-Maxwell binary diffusion coefficient. Alternatively, we may introduce molar quantities and the frictional coefficient ζ_{ij} [7]

$$\frac{\partial \mu_i}{\partial x} = \Sigma \, \zeta_{ij} \, m_j (v_j - v_i)$$

The relationships between all forms of the frictional coefficients are

$$\frac{RTx_j}{Ð_{ij}} = \zeta_{ij} \, m_j = f_{ij}$$

and more generally

$$(J_i)_v = - \frac{m_i}{f_{ij}\left(1 + \dfrac{m_i V_i}{m_j V_j}\right)} \frac{\partial \mu_i}{\partial x} \tag{13}$$

It is seen in Eqs. (11) and (12) that the diffusion coefficients are superficially comprised of two factors; a frictional term as represented by f_{12} or f_{21} and a thermodynamic term $\partial\mu_1/\partial m_1$ or $\partial\mu_2/\partial m_2$. However, some caution should be levelled at this description because the two terms are closely connected as seen by Eqs. (8) and (9) which describe the direct relationship between the gradient of the chemical potential and the frictional term.

The quantity $(\partial\mu_2/\partial m_2)_{T,P}$ can be expressed in terms of thermodynamic non-ideality coefficients A_2 and A_3 [8] such that, when combined with Eq. (12), gives

$$(D_{22})_v = RT \frac{\bar{M}_{2,n}}{f_{21}} \left(\frac{1}{\bar{M}_{2;n}} + 2A_2 C_2 + 3A_3 C_2^2 + ...\right) \tag{14}$$

where $\bar{M}_{2,n}$ is the number average molecular weight of component 2 and C_2 the concentration units of M, g m^{-3}.

There are a number of quantitative features of Eq. (14) which are important in relation to rapid diffusional transport in binary systems. The mutual diffusion coefficient is primarily dependent on four parameters, namely the frictional coefficient f_{21}, the virial coefficients, molecular weight of component 2 and its concentration. Therefore, for polymers for which water is a good solvent (strongly positive values of the virial coefficients), the magnitude of $(D_{22})_v$ and its concentration dependence will be a compromise between the increasing magnitude of f_{21} with concentration and the increasing value of the virial expansion with concentration.

2.2 Mutual Diffusion of Polymers

2.2.1 Neutral Polymers

We have used the uncharged polysaccharide dextran as a model describing the behaviour of water-soluble polymers. The dextrans used in this study have about 95 % α-(1 $\rightarrow$ 6) linkages within the main chain and side chains; the 5 % non-α-(1 $\rightarrow$ 6) linkages are starting points of branched chains of which most are only 'stubs' of about two glucose units [9]. Therefore, while there is some branching in dextran, albeit low, its solution behaviour is that of a linear, random-coil molecule [10,11].

The measurement of the mutual diffusion coefficient of dextrans of differing molecular weights (Fig. 1) allows us to evaluate the interplay of the various terms constituting $(D_{22})_v$ as given in Eq. (14). The diffusion coefficients have been measured in solutions covering a concentration range from 1 to 250 kg m^{-3} and for dextrans over a molecular weight range from 10^4 to 2×10^6. Dextrans with molecular weights $\bar{M}_w > 7 \times 10^4$ exhibit a marked concentration dependence of $(D_{22})_v$; with increasing concentration the mutual diffusion coefficient becomes greater. On the other hand,

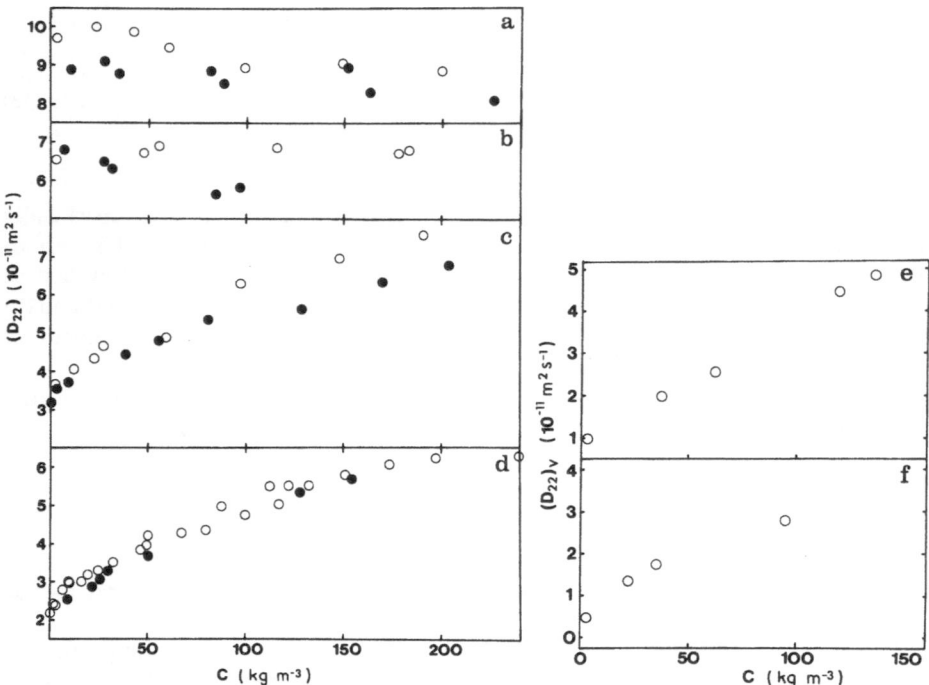

Fig. 1 a–f. The mutual diffusion coefficient $(D_{22})_v$ of dextran as a function of dextran concentration for **a** dextran T10 ($\bar{M}_w \sim 10^4$), **b** dextran T20 ($\bar{M}_w \sim 2 \times 10^4$), **c** dextran T70 ($\bar{M}_w \sim 7 \times 10^4$), **d** dextran FDR7783 ($\bar{M}_w \sim 1.5 \times 10^5$), **e** dextran T500 ($\bar{M}_w \sim 5 \times 10^5$), and **f** dextran T2000 ($\bar{M}_w \sim 2 \times 10^6$): 0, values of D_{22} obtained by measurement of the by concentration gradient relaxation as monitored by refractive index methods; (●), values of D_{22} obtained by photon correlation spectroscopy. Data obtained from ref. [1-] and unpublished work. For earlier studies of dextran mutual diffusion in concentrated solutions see also ref. [13]

for dextrans with $\bar{M}_w = (1 \text{ to } 2) \times 10^4$ the mutual diffusion coefficient is essentially constant and decreases for low molecular weight samples. At high dextran concentrations, the diffusion coefficient appears to converge to a constant value, in the range of $(6-8) \times 10^{-11}$ m^2 s^{-1} irrespective of the dextran molecular weight.

On closer analysis of the individual parameters in Eq. (14) we find that the virial expansion is a parameter essentially independent of molecular weight at high dextran concentrations (Fig. 2). Therefore, it is recognized that the M/f_{21} term in Eq. (14) must become a molecular weight-independent parameter.

---

2 Diffusion in a binary system may also be determined by measurement of the intradiffusion coefficient (sometimes referred to as the self-diffusion coefficient), D_i^+. In the case of intradiffusion, no net flux of the bulk diffusant occurs; the molecules undergo an exchange process. Measurements are usually carried out by using trace amounts of labelled components in a system free of any gradients in the chemical potential. The molecular movement of the solute is governed by frictional interactions between labelled solute and solvent, and labelled solute and unlabelled solute.

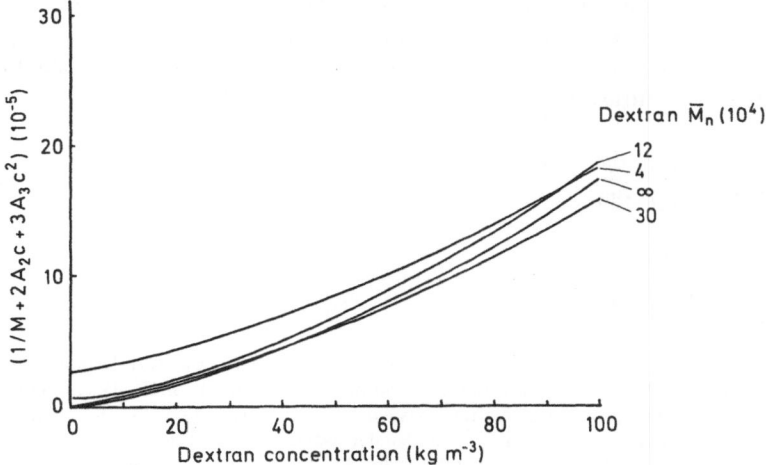

Fig. 2. Calculated values of the virial expansions in Eq. (14), i.e. $\left(\dfrac{1}{\overline{M}_{2,n}} + 2A_2C_2 + 3A_3C_2^2 + \ldots\right)$ as a function of dextran molecular weight. The virial coefficients and molecular weights are summarised in ref. [12]. Virial coefficients for cross-linked dextran ($\overline{M}_n \to \infty$), i.e. 'Sephadex' beads, have been obtained from ref. [14]

Evidence that this is the case has also been obtained from sedimentation velocity studies on dextrans where the sedimentation coefficient S, which is proportional to M/f_{21}, is shown to be molecular weight-independent at high concentrations [15]. Furthermore, studies on the intradiffusion of tritiated water (HTO) in dextran solution [8] (footnote 2) revealed that the term

$$\frac{M}{f_{21}} \propto \frac{(D_1^+)^\circ \, D_1^+ \, m_2}{[(D_1^+)^\circ - D_1^+]m_1},$$

in which D_1^+ and $(D_1^+)^\circ$ are the intradiffusion coefficients of HTO in dextran solutions and in water respectively, is not only molecular weight-independent but concentration-independent as well up to dextran concentrations of 250 kg m^{-3} [8]. These results point to the possibility that the molecular weight independent parameter (at high concentrations), M/f_{21}, may be related to the segmental properties of the dextran in concentrated solutions rather than to a description of individual molecules. Unfortunately, an independent estimate of f_{21} and therefore the effective value of M cannot be made using this phenomenological approach (footnote 3).

Another approach in understanding the behaviour and structure of concentrated solutions is offered by the dynamic and static properties of the network models of

---

3 Previously, the assumption has been made that the polymer self-diffusion coefficient is proportional to $1/f_{21}$. The assumptions involved in the derivation of this relationship have been shown experimentally to be invalid [8].

semi-dilute polymer solutions (footnote 4) derived by De Gennes and others (Ref. [16] and references cited therein). The co-operative motions of the chains corresponding to long wavelength functions of polymer concentration have been described in terms of a co-operative diffusion coefficient, D_{coop}, which is expressed as

$$D_{coop} \simeq \frac{M(\xi)}{f(\xi)} \frac{\partial \mu}{\partial m}$$

$$= \frac{kT}{6\Pi \eta_0 \xi}$$

---

4 There is general agreement that polymer solutions exhibit distinct features due to intermolecular interactions between the polymer molecules. In dilute solutions, of course, the molecules are independent of one another. The physical properties of such molecules are defined by the intrinsic properties of the individual molecules such as molecular weight. When the concentration is increased intermolecular interactions will gradually occur. At a certain concentration, defined as the critical concentration C*, the domains of the polymer molecules will be in a state of frequent contact with their neighbours. In this regime, the physical properties of the molecules depend on concentration and become essentially independent of molecular weight.

The structure and dynamic properties of this interacting array of polymer molecules depend on polymer conformation and flexibility. This also applies to the magnitude of C*, which can be calculated on the basis of various geometric packing arrangements of the characteristic domain of the molecules.

Polymer conformations and domains may be considered over a broad range. At one end of the spectrum we consider the relatively rigid, compact polymer, with high segment density and low content of micromolecular solvent (water). Such molecules could be regarded as impenetrable spheres, e.g. globular proteins such as albumin. Similar considerations could also apply to smaller molecules of limited flexibility. Surface contacts between these molecules dominate intermolecular interactions. However, little is known of the nature of these interactions occurring in concentrated solutions. At the other end of the spectrum of polymer conformations, we consider the behaviour of rigid or semirigid rod-like molecules. These solutions are characterised by a high viscosity at relatively low polymer concentrations. The interacting array of rods may be compared with a random pile of match sticks. This network of a random array of rods forms a much more porous structure, containing relatively high concentrations of micromolecular solvent, than that envisaged for interacting compact spheres. There are only few experimental studies on the properties of interacting rods. Generally, at moderate concentrations these systems are unstable in solution and transfer into a tactoidol or an aggregate phase, e.g. collagen and tropomyosin.

Between these two extremes offered by rods and spheres lie the networks formed by molecules of varying flexibility and configuration.

To determine the structure of these different types of networks we have extensively studied (Ref. [1] and unpublished work) the diffusional behaviour of materials within these concentrated solutions. These diffusants are used in trace quantities so that they do not perturb the physicochemical properties of the network. By varying the size of the diffusional probe a good deal of information can be obtained about the structure of the network due to the hindrance effect the latter exerts on the diffusant.

The major theoretical interest in the structure of polymer solutions has centered on the behaviour of linear polymers exhibiting considerable flexibility and conformation in solution. Due to their chain-like nature these polymer molecules assume an extended coil-like configuration. Their behaviour in concentrated solutions has been the subject of extensive theoretical work (see De Gennes [16] and references cited therein). The basis of these theories is that at concentrations higher than that at which overlapping of polymer domains occur, the motion of each polymer chain is governed by some form of entanglement. The solution is now viewed as a transient network with a characteristic elastic modulus and dynamic properties governed by the gel-like character of the solution; the network having a characteristic mesh size ξ which represents the average

where the term $M(\xi)/f(\xi)$ describes the properties of a blob of length ξ; k is Boltzman's constant, T the temperature and η_0 the solvent viscosity. The diffusion coefficient D_{coop} in semi-dilute solution is independent of molecular weight and scales with concentration such that

$$D_{coop} \sim C^{-\gamma} \tag{15}$$

where $\gamma = 0.75$ for polymer diffusion in good solvents.

distance between adjacent contact points. This fundamental parameter depends on the polymer concentration but is independent of the molecular weight of the solute. The network composed of physically entangled macromolecules is usually visualized as a sequence of blobs of size ξ, each blob occupying a volume proportional to ξ^3. The term 'blob' denotes the chain portion between successive entanglement points. Therefore, at concentrations equal to or higher than the critical concentration C^* at which overlapping of coils occurs (note that $C=C^*$ denotes a closely packed system of non-overlapping coils), the 'blob' model of a polymer chain is invoked, and it is at this concentration that divergence from dilute solution behaviour is anticipated. Each blob acts as an individual unit with both hydrodynamic and excluded volume interactions; for blobs of the same chain, all interactions are screened out. We have established an important correlation between the parameter C^* and the onset of rapid polymer transport (see 3.3.3.).

The molecular weight-independent properties of flexible chain networks were early recognised through the studies of the sedimentation behaviour of polysaccharides (dextran) as a function of concentration [15]. Dextran molecules of different molecular weights sediment at very different rates at high dilution. However, at concentrations above a critical concentration they all move at a relatively low rate. This is in agreement with the prediction of the dynamic behaviour of coiled molecules which entangle above the critical concentration; their dynamic behaviour is independent of molecular weight. Above the critical concentration, the whole polysaccharide network moves as a plug through the ultracentrifuge cells. Similar results were obtained for the sedimentation of different molecular weight fractions of the anionic polysaccharide, hyaluronate [17]. The only other main evidence for entanglement comes from rheological data [18]. Aharoni [19] has detected several different levels of organisation of the so-called 'entanglements' or interacting units by dynamic (viscosity) and static (light scattering) measurements. With increasing concentration he found a critical concentration which determines the onset of intermolecular interpenetration in the case of complete space-filling without change in the intramolecular segmental density. The next stage is reflected by the appearance of dynamically measurable entanglements in the peripheral regions of the molecular coils and finally, at a high concentrations, by the appearance of such entanglements in the core regions or the coils. Certainly such interpenetration and entanglements are only exhibited by flexible molecules and not by rigid rods and spheres.

The structure of concentrated solutions of branched molecules has also received little attention. It is probable that the network formed by entangled branched molecules displays a topological structure which is different from that formed by linear chains. It is known that the sedimentation properties of branched synthetic polymers differ, especially in good solvents, from those of linear polymers. The concentration dependence of the sedimentation coefficient is relatively more pronounced for branched polymers than that for linear ones [20].

It should be noted that in studies cited above potential changes in the molecular dimensions of polymers with increasing concentration were not reported. Early studies by Maron et al. [21] and more recently by Ogston and Preston [22] have shown that flexible coiled macromolecules may occupy a reduced volume in concentrated polymer solution. They have identified this phenomenon with a collapse of the molecule to a more compact coil, with a concomitant increase in intramolecular segmental density. They demonstrated that the viscosities of concentrated dextran solutions may be interpreted in terms of the osmotic shrinkage of the molecular domain of the polymer due to the high osmotic pressure of its environment, i.e. neighbouring molecules compete for water with the result that water is withdrawn from the fully hydrated domain of the molecule. Ultimately, polymer contraction or expansion with increasing concentration will be determined by a number of different types of interactions including segment/solvent interaction and both intra- and inter-

Quasi-elastic light scattering (QELS) technique has been widely used for polymer diffusion in non-aqueous solvents to measure D_{coop}. Diffusion coefficients obtained by this non-perturbing technique, which monitors the relaxation of the random concentration gradients produced by the thermal fluctuations in solution, have been found to be in good agreement with the time-independent mutual diffusion coefficient corresponding to the relaxation of an applied concentration gradient [25]. Such correspondence has also been established by systematic studies of polymer diffusion in aqueous solution, namely dextran diffusion [12].

The scaling treatment, as described by Eq. (15) for D_{coop}, has received general interest in assessing experimental data in non-aqueous systems for semidilute polymer solutions (for a review see ref. [20]). However, experimental data of dextran diffusion in water, while converging to molecular weight independence at high dextran concentrations, are not compatible with the scaling law predictions of Eq. (15) which yields an exponent of 0.75. Rather, we have found exponents only as high as 0.3 for the highest molecular weight fractions [12]. Several reasons have been reported for small exponents:

(i) the scaling relations only emerge from asymptotic dimensional analysis assuming infinitely long chains,
(ii) the polymer is not sufficiently flexible,
(iii) the solvent is not 'good' enough.

Therefore, while the independence of dextran diffusion on molecular weight is approximately established at high concentrations, the applicability of Eq. (15) in expressing D_{coop} directly as a function of C is in doubt in this case since the predictions of the concentration dependence of D_{coop} are not upheld with this material.

2.2.2. Polyelectrolytes

We have discussed above neutral polymers such as dextrans whose virial coefficients, which arise primarily from excluded-volume effects, yield relatively high diffusion coefficients for high molecular weight fractions. When introducing electrically positive or negative moieties into the polymer chain, polyion-microion interactions and the Donnan equilibrium must also be taken into account in the virial term. For good solvents, the Donnan term will dominate the excluded-volume term in the virial expansion, thus giving rise to relatively high values of A_2, A_3 etc. and introducing a marked sensitivity of these coefficients to the ionic strength of the solution. Such considerations have been demonstrated in studies of the mutual diffusion of the anionic polysaccharide hyaluronate (as measured by absorption optics using the light

---

segment/segment interactions between molecules and the dependence of these interactions on the segmental density distribution of each polymer molecule. These interactions encompass entropic effects as embodied by excluded-volume terms and enthalpic effects associated with binding and aggregation.

Indeed, at very high polymer concentrations enhancement of these effects occurs. Recent studies by Franks et al. [23] on the rheological behaviour and "freeze fracture" electron microscopical analysis of several synthetic linear flexible polymers, including poly(vinylpyrrolidone) and poly-(ethylene glycol) in concentrated solutions, suggest that these molecules do not form a network mesh but rather exhibit aggregation. Anionic polysaccharides, on the other hand, are known to form an anisotropic packing array in condensed films. These films may be stretched to enhance orientation and be used for X-ray diffraction studies [24].

absorption of a covalently bound fluorescein group and Raleigh interference optics in the analytical ultracentrifuge) [26]. The concentration dependence of the mutual diffusion coefficient of the electroneutral hyaluronate ($\overline{M}_w \simeq 10^6$) is considerably larger than that observed at similar concentrations of the uncharged polysaccharide dextran. The increase in mutual diffusion with decreasing ionic strength is a reflection of the marked increase in the thermodynamic virial coefficients with decreasing ionic strength. We have obtained mutual diffusion coefficients as high as 50 times the normal rate of infinite dilution. The largest diffusion coefficient recorded (30×10^{-11} m^2 s^{-1}) is equivalent to that of a pentasaccharide in free solution. The rapid transport of the electroneutral polymer may be considered as resulting from the balance between the relatively rapid mobility of the counterion and the slow moving polyanion. The requirement for electronentrality causes a retardation of the migration velocity of the counterion but an increase in the transport of the polymer component. These relative mobilities and transient charge separations will be strongly influenced by the screening effect of the ambient simple electrolyte. Note also that in the studies of polyelectrolyte diffusion in simple electrolyte-water mixtures we are dealing with a multicomponent system. Later, we shall discuss that certain multi-component systems may exhibit a transition from normal diffusion to an alternative transport regime involving ordered convective flows as a result of interacting flows between components (see also Refs. [27] and [28]). While this transition is difficult to detect by classical transport techniques it does lead to rapid polymer transport. Such effects have not been discounted in the transport studies of the hyaluronate discussed above.

2.3 Solvent Diffusion in Polymer Solutions

We have already noted vide supra in Eq. (5) that in a volume-fixed frame of reference the mutual diffusion coefficient of the micromolecular solvent, H$_2$O, has the same value as the polymer diffusion coefficient, i.e. the diffusion coefficient describing the boundary relaxation of the polymer concentration gradient is exactly the same as that coefficient describing the relaxation of the corresponding opposite concentration gradient of the solvent. These diffusion coefficients, which lie in the range of 0.01 to 10×10^{-11} m^2 s^{-1}, are considerably lower than the self-diffusive mobility of water measured, for example, by intradiffusion of HTO in systems where there are no bulk concentration gradients, thus, the value $\simeq 200 \times 10^{-11}$ m^2 s^{-1} at 25 °C is obtained [8].

In a volume-fixed reference plane we can then conveniently speak of an equal exchange of volume flows of the two components in a binary system across a plane. The magnitude of the flow, as described by Eq. (13), is directly related to the chemical potential gradient or, more commonly, to the osmotic pressure (Π) gradient

$$\frac{\partial \Pi}{\partial x} \simeq - \frac{\partial \mu}{\partial x} \tag{16}$$

Therefore, the volume flow of water, corresponding the volume exchange diffusional flux with polymer solute, can be regarded as osmotic flow caused by an osmotic pressure gradient [29]. The magnitude of this flow does however not effect the magnitude of the mutual diffusion coefficient as the latter is normally a differential diffusion coefficient which is independent of the concentration gradient. In the case of dextran, we have found that its mutual diffusion coefficient is independent of the osmotic pressure gradient operating across the boundary at a definite mean concentration of the dextran at the boundary [30].

The flow of water under the conditions of rapid solute diffusion in semidilute solutions provokes considerable interest in the kinetics of the swelling behaviour of polymer gels. There is growing evidence that the dynamic diffusion behaviour of a polymer in the form of either a gel or in semidilute solution is similar. Tanaka and Fillmore [31] have examined a set of spherical polyacrylamide/water gels and measured their characteristic swelling times from which they calculated the diffusion coefficient of polyacrylamide chains. From the same samples at swelling equilibrium, the diffusion coefficient was also obtained by QELS experiments. The two methods were in good agreement. These experiments showed that the swelling kinetics is controlled by the diffusion coefficient of the polymer network. Furthermore, Munch et al. [32] have studied the diffusion controlling the concentration fluctuations in semidilute solutions of polystyrene and in polystyrene gels by QELS experiments and found them similar (see also review [33]). These experiments then point to possible similarities between the diffusion behaviour of semidilute dextran solutions and the swelling of cross-linked dextran gels (commercially supplied under the name of Sephadex from Pharmacia, Sweden) [49].

3 Polymer Transport in Ternary Systems

3.1 Basic Theoretical Considerations

Studies on multicomponent systems have been mainly restricted to relatively simple ternary systems containing a solvent as component 1 and solutes as components 2 and 3. For such a system, under zero-volume flow conditions (Eq. (3)), exact expressions for the fluxes of the components 2 and 3 may be written as independent quantities of the forces involved so that the linear laws to which the Onsager reciprocal relationship applies may be written as follows [34]:

$$J_2 = L_{22}Y_2 + L_{23}Y_3 \tag{17}$$

$$J_3 = L_{32}Y_2 + L_{33}Y_3 \tag{18}$$

where

$$Y_i = -\sum_{j=2}^{3} \left(\delta_{ij} + \frac{m_j V_i}{m_1 V_1} \right) \frac{\partial \mu_i}{\partial x} \tag{19}$$

δ_{ij} is the Kronecker delta and L_{ij} are the phenomenological coefficients such that $L_{23} = L_{32}$ which is in accordance with the Onsager reciprocal relationship. Rewriting the system in terms of composition variables and ternary diffusion coefficients D_{ij} gives

$$J_2 = -D_{22} \frac{\partial m_2}{\partial x} - D_{23} \frac{\partial m_3}{\partial x} \tag{21}$$

$$J_3 = -D_{32} \frac{\partial m_2}{\partial x} - D_{33} \frac{\partial m_3}{\partial x} \tag{22}$$

D_{ii} are the principal or main diffusion coefficients and D_{ij} the interaction or cross-diffusion coefficients. These diffusion coefficients can be expressed by phenomenological coefficients and chemical potentials as follows:

$$- D_{22} = aL_{22} + bL_{23} \tag{23}$$

$$- D_{23} = cL_{22} + dL_{23} \tag{24}$$

$$- D_{32} = aL_{32} + bL_{33} \tag{25}$$

$$- D_{33} = cL_{32} + dL_{33} \tag{26}$$

where the expressions for a, b, c and d are

$$a = \left[\left(1 + \frac{m_2 V_2}{m_1 V_1}\right) \mu_{22} + \frac{m_3 V_2}{m_1 V_1} \mu_{32} \right] \tag{27}$$

$$b = \left[\frac{m_2 V_3}{m_1 V_1} \mu_{22} + \left(1 + \frac{m_3 V_3}{m_1 V_1}\right) \mu_{32} \right] \tag{28}$$

$$c = \left[\left(1 + \frac{m_2 V_2}{m_1 V_1}\right) \mu_{23} + \frac{m_3 V_2}{m_1 V_1} \mu_{33} \right] \tag{29}$$

$$d = \left[\frac{m_2 V_3}{m_1 V_1} \mu_{23} + \left(1 + \frac{m_3 V_3}{m_1 V_1}\right) \mu_{33} \right] \tag{30}$$

with $\mu_{ij} = \partial\mu_i/\partial m_j$
The expressions for L_{ij} in terms of frictional coefficients read:

$$L_{22} = m_2 \left\{ \left[1 + \frac{m_3 V_3}{m_1 V_1}\right] \left[f_{31} \left(1 + \frac{m_3 V_3}{m_1 V_1}\right) + f_{32}\right] \right.$$
$$\left. + \frac{m_2 V_3}{m_1 V_1} \left[f_{21} \frac{m_3 V_3}{m_1 V_1} - f_{23}\right] \right\} \Big/ Q \tag{31}$$

$$L_{33} = m_3 \left\{ \left[1 + \frac{m_2 V_2}{m_3 V_3}\right] \left[f_{21} \left(1 + \frac{m_2 V_2}{m_1 V_1}\right) + f_{23}\right] \right.$$
$$\left. + \frac{m_3 V_2}{m_1 V_1} \left[f_{31} \frac{m_2 V_2}{m_1 V_1} - f_{32}\right] \right\} \Big/ Q \tag{32}$$

$$L_{32} = L_{23} = -m_2 \left\{ \left[f_{21} \frac{m_3 V_3}{m_1 V_1} - f_{23} \right] \left[1 + \frac{m_2 V_2}{m_1 V_1} \right] \right.$$

$$\left. + \frac{m_3 V_2}{m_1 V_1} \left[f_{31} \left(1 + \frac{m_3 V_3}{m_1 V_1} \right) + f_{32} \right] \right\} \Big/ Q$$

$$= -m_3 \left\{ \left[f_{31} \frac{m_2 V_2}{m_3 V_3} - f_{32} \right] \left[1 + \frac{m_3 V_3}{m_1 V_1} \right] \right.$$

$$\left. + \frac{m_2 V_3}{m_1 V_1} \left[f_{21} \left(1 + \frac{m_2 V_2}{m_1 V_1} \right) + f_{23} \right] \right\} \Big/ Q \tag{33}$$

$$\text{where } Q = \left[\left[1 + \frac{m_2 V_2}{m_1 V_1} \right] \left[1 + \frac{m_3 m_3}{m_1 V_1} \right] - \frac{m_2 m_3 V_2 V_3}{(m_1 V_1)^2} \right]$$

$$\times \left[\left[f_{31} \frac{m_2 V_2}{m_1 V_1} - f_{32} \right] \left[f_{21} \frac{m_3 V_3}{m_1 V_1} - f_{23} \right] \right.$$

$$\left. - \left[f_{21} \left(1 + \frac{m_2 V_2}{m_1 V_1} \right) + f_{23} \right] \left[f_{31} \left(1 + \frac{m_3 V_3}{m_1 V_1} \right) + f_{32} \right] \right] \tag{34}$$

In comparison with the qualitative description of diffusion in a binary system as embodied by Eqs. (11), (12) or (14), the thermodynamic factors are now represented by the quantities a, b, c, and d and the dynamic factors by the phenomenological coefficients which are complex functions of the binary frictional coefficients. Experimental measurements of D_{ij} in a ternary system, made on the basis of the knowledge of the concentration gradients of each component and by use of Eqs. (21) and (22), have been reviewed [35]. Another method, which has been used recently [36], requires the evaluation of μ_{ij} from thermodynamic measurements such as osmotic pressure and evaluation of all f_{ij} from diffusion measurements and substitution of these terms into Eqs. (23)–(26).

3.2 Component Distributions in a Ternary System

Several experimental systems can be set up corresponding to different initial component distributions as schematically shown in Fig. 3; we shall confine our considerations to the transport arising from relaxation of the concentration gradients of only one or both solute components in a system. System A: one of the components is initially present at uniform concentration throughout the system while the other component is maintained at a non-zero concentration gradient. This system has been a focus of major study in our laboratory and will be discussed in Section 3.3

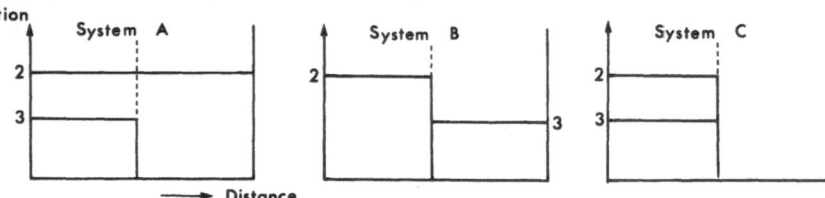

Fig. 3. Schematic representation of various initial component distributions that can be studied in ternary diffusion

at some length in terms of rapid polymer transport. System B: the two solute components have opposite concentration gradients and interdiffuse. This system has been extensively considered (Sect. 5.4) in terms of low molecular weight solutes and heat transfer. It is occasionally described under the heading 'double diffusive convecting system'. System C: the two components have concentration gradients in the same direction. This system has to be studied systematically and will not be discussed here. It will probably be of interest in the future with respect to its potential use in measuring ternary diffusion coefficients using radiolabelled species.

3.3 Transport in Polymer/Polymer/Solvent Mixtures

3.3.1 Historical Perspective

Some early studies on the diffusion of globular proteins in polysaccharide solutions [37,38] (system A) revealed that the movement of the proteins is markedly retarded. The diffusive retardation of the globular proteins may be described by the empirical relationship

$$D/(D)^\circ = A^* \exp\left(-1.4\,r.\,L^{0.5}\right) \tag{35}$$

where D is the diffusion coefficient of the protein in the polysaccharide solution, $(D)^\circ$ the diffusion coefficient of the protein in the absence of the polysaccharide, A^* a constant close to unity, r the radius of the diffusing particle, and L the polysaccharide concentration expressed as cm polymer per cm^3 solution.

The retardation of the protein movement has been discussed qualitatively in terms of a sieving mechanism rather than a frictional resistance [37]. Ogston et al. [39] have theoretically described the diffusion as a stochastic process in which the particles move by unit displacements and in which the decrease in the rate of diffusion in a polymer network depends on the probability that a particle finds a "hole" in the network into which it can move. The relationship derived from this approach is in close agreement with Eq. (35).

The importance of the shape of the diffusing molecule was then investigated in these systems by the analysis of the diffusion of various asymmetric particles in hyaluronate networks [40]. These systematic studies were performed at low concentration gradients of the diffusing macromolecules. The diffusion of various linear polymers, including DNA and polyadenylate, in dilute hyaluronate solutions, was measured. The empirical relation (Eq. (35)) describing globular particle diffusion holds for the diffusion of these linear polymers [45]. However, estimates of the effective radii of the asymmetric solutes in the polysaccharide networks by a similar equation derived by Ogston et al. [39] have revealed that linear polymers behave as if they had smaller dimensions than those measured in free solution. The most striking effect was obtained with DNA. Although its equivalent hydrodynamic radius in free solution is of the order of 25—30 nm, its diffusive retardation in the polysaccharide solution is the same as that of a globular particle with a radius of ~ 8 nm. The authors concluded that the asymmetric particles may have moved end-on in the network since this movement should encounter less obstacles than a random movement of the whole hydrodynamic unit. That a linear chain molecule may perform such a movement (reptation) had been earlier suggested by De Gennes (see Ref. [16]).

In a recent investigation on the diffusion of various macromolecules across a membrane consisting of a cartilage matrix deposited in tissue culture, Cumming et al. [41] have shown that globular proteins are retarded as expected. However two asymmetric molecules, fibrinogen and collagen, are transported faster than their diffusional behaviour in free solution would predict. Such effects could be explained if the molecules were forced into end-on movements since a rod-like particle would encounter less resistance moving lengthwise than when moving broadside.

While the studies described above have demonstrated a retarding effect of dilute polysaccharide solutions on the diffusion of various macromolecules (which were initially present at low concentrations), an effect of the diffusant on the polysaccharide solution has not been detected so far. To our knowledge, the only attempt at determining quantitatively the transport of both polymers in a ternary system is that of Cussler and Lightfoot [42] who studied the ternary diffusion of two polystyrenes of different molecular weights in toluene. In this case, the non-zero cross diffusion coefficients increase strongly with concentration and are of a magnitude similar to the principal diffusion coefficients at high concentrations.

In progression to transport studies in concentrated polymer mixtures, several surprising features were early observed. It has been shown [43] in a limited number of experiments in the analytical ultracentrifuge that the transport of linear, flexible macromolecules within concentrated polymer solutions (with component distribution corresponding to System A) proceeds at much higher rates than those observed in the absence of the polymer. Further experiments performed by Preston and Kitchen [44], using radiolabelled polymers and an open-ended capillary technique, revealed that the rapid transport appears to be a general phenomenon associated with the movement of a range of linear, flexible macromolecules but not with globular particles. On the contrary, globular proteins, for example, showed a decrease in their transport rates in concentrated polymer solutions as compared to their value in the absence of the polymer. Furthermore, a marked time-dependence of the polymer transport coefficients [45] was found using the open-ended capillary method. In contrast, the behaviour of solvent markers, such as [^{14}C]sorbitol, is ideal and exhibits no time dependence.

The rapid transport of the linear, flexible polymer was found to be markedly dependent on the concentration of the second polymer. While no systematic studies were performed on these ternary systems, it was argued that the rapid rates of transport could be understood in terms of the dominance of strong thermodynamic interactions between polymer components overcoming the effect of frictional interactions; this would give rise to increasing apparent diffusion coefficients with concentration [28,45]. This is analogous to the resulting interplay of these parameters associated with binary diffusion of polymers.

In view of the highly 'unusual' nature of these results and the lack of a routine method for transport measurements unambiguously establishing that rapid transport was indeed a real manifestation of the system, our studies on rapid polymer transport remained unreported in detail. However, in a recent article [46] we have demonstrated that rapid polymer transport actually occurs in these systems due to the formation of ordered macroscopic structures which move rapidly. This rapid transport has been shown to be not the result of bulk convection since normal diffusional kinetics was observed for solvent markers such as [^{14}C]sorbitol. The striking feature of this new type of transport process is that it is accompanied by ordered structured flows in the form of finger-like projections emanating in both directions perpendicular to the initial boundary as observed by tagging the polymers with coloured dye. The following sections give a review of our extensive work on these structured flow systems.

3.3.2 Experiments on Rapid Polymer Transport Carried Out on a Standard System

To study the nature of this rapid polymer transport in detail, this section will be concerned with a series of experimental measurements on one particular system, namely a solution of dextran T10 ($\bar{M}_w \simeq 10^4$) with a uniform concentration of 135 kg m^{-3} and an imposed gradient of a linear, flexible poly(vinylpyrrolidone) ($\bar{M}_n \simeq 3 \times 10^5$) (PVP 360). This gradient initially extended from 5 kg m^{-3} to zero concentration. The choice of using the polymers at this concentration was based upon our earlier work [44] in which it was shown that nearly maximal transport rates of PVP 360 occur in such a system. This system will be referred to as the standard system. The phase diagram of this PVP 360/dextran T10 mixture clearly demonstrates that the transport experiments were performed within the one-phase region [47].

3.3.2.1 Sundelöf Transport Cell

A detailed kinetic analysis of the rapid transport of PVP 360 in the solution of dextran has been made with the aid of a newly developed diffusion cell [48] in which horizontal boundaries are formed by a shearing mechanism in a cylinder 5 mm in diameter and 10 mm high (Fig. 4). The transport of radiolabelled [³H]PVP 360 over the boundary does not follow normal diffusional kinetics, i.e. it is not linear with the square root of time (Fig. 5). Instead, the transport appears to be linear with time up to about 3 h; then, it levels off. Within the first 3 h, about 70–80% of the PVP

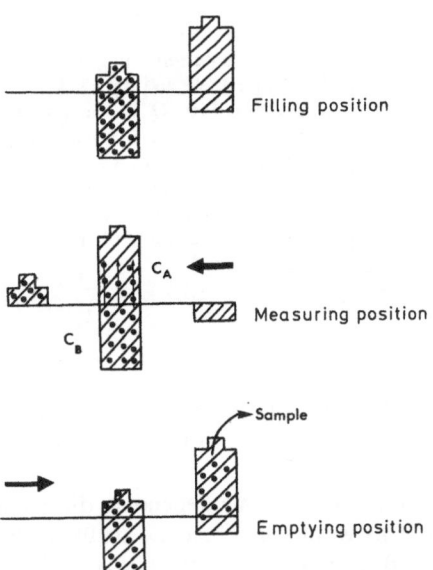

Fig. 4. Schematic representation of the formation of the free liquid boundary in the Sundelöf diffusion cell. C_B and C_A represent the concentration of the studied material in the bottom and top chambers of the cell, respectively

concentration gradient has relaxed, the linear rate was measured as 1.4×10^{-3} m h^{-1}. We have also shown that the transport of trace quantities of [^{14}C]sorbitol, used as a solvent marker, follows normal diffusion kinetics and is much slower than the PVP transport [46,47,50]. This result suggests that there is no massive bulk convection occurring in the system. We have also demonstrated that the kinetics of the dextran T10 transport, as measured by both forward and back fluxes of [^{3}H]dextran T10, is similar to that of the PVP 360. However, the resultant net flux of dextran across the boundary is essentially zero [46,50].

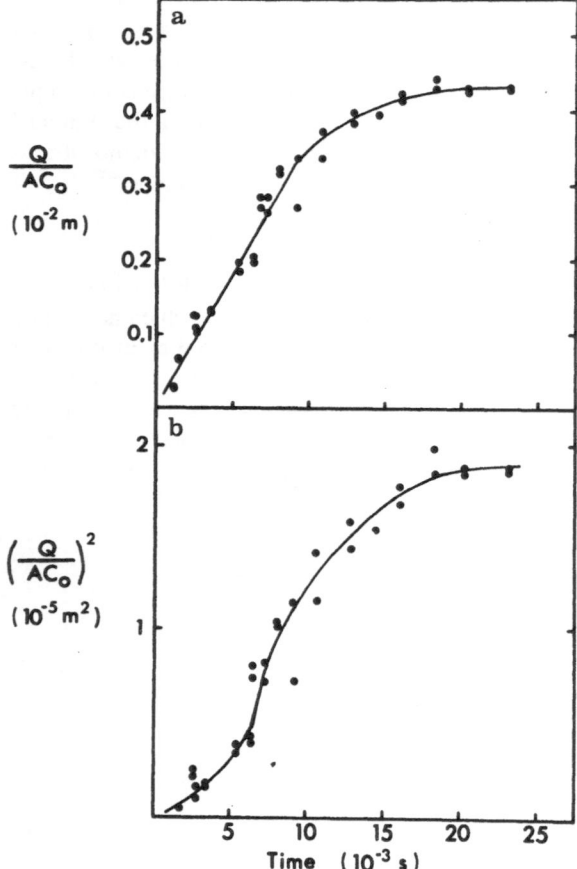

Fig. 5a and b. Transport of [^{3}H]-PVP over a boundary formed by layering 135 kg m^{-3} dextran T10 onto a solution containing 5 kg m^{-3} [^{3}H]PVP 360 and 135 kg m^{-3} dextran T10. Q is the amount transported over a boundary of surface area A and C_0 the initial concentration in the lower compartment. **a** plot of (Q/AC_0) versus time, **b** plot of $(Q/AC_0)^2$ versus time [47]

The overall significance of describing the relaxation of the PVP gradient as a linear response with time and the associated change from fast to slow transport in the kinetic analysis is not obvious.

Alternatively, we have described the system in terms of a time-dependent diffusion coefficient from the data presented in Fig. 5. This would also be conceivable if the diffusion coefficients described for ternary diffusion in Eqs. (21) and (22) were markedly concentration-dependent; only their magnitude appears to be abnormally high.

3.3.2.2 Open-Ended Capillary Technique

An analysis of the PVP transport using open-ended capillary technique (with capillary dimensions 0.92 mm in diameter and 10 mm high) has in fact clearly demonstrated the time dependence of the apparent diffusion coefficient of the [^{3}H]PVP 360 transport (Fig. 6). On the other hand, the transport behaviour of trace quantities

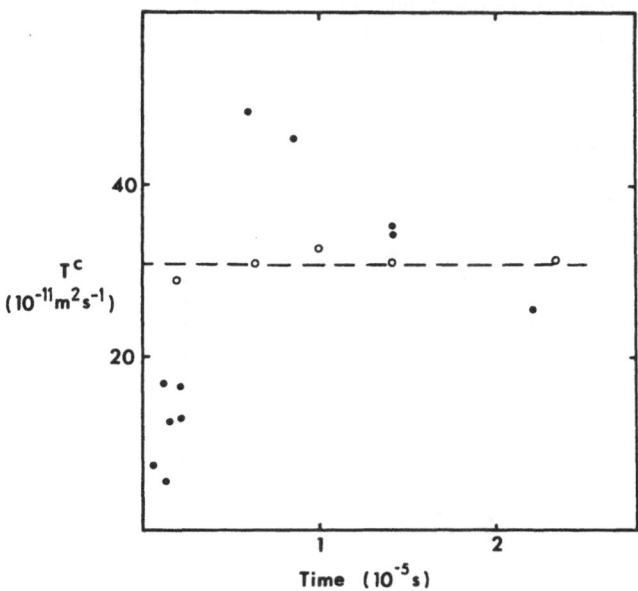

Fig. 6. Variation of T^c with time (transport coefficient obtained from diffusion analysis of results from open-ended capillary) for the [^{3}H]PVP 360 (●) and [^{14}C]sorbitol (○) transport in standard solution systems (described in the legend to Fig. 5)[47]

of [^{14}C]sorbitol is ideal in that it exhibits no time dependence. The mass transfer for [^{3}H]PVP in the capillaries can be analysed in the same fashion as those obtained by use of the Sundelöf cells, and a very similar kinetic behaviour has been obtained in both cases[47]. The only difference is that, at a dextran concentration of 135 kg m^{-3}, the PVP transport occurs considerably more rapidly in the Sundelöf cell than in the capillary.

3.3.2.3 Transport Analysis Using the Ultracentrifuge

Schlieren Optics

Since this optical system measures changes in the refractive index gradients in the solutions arising from concentration gradients of any of the solutes, standard diffusion analysis would be subject to errors if the PVP concentration gradients ultimately induced concentration gradients of dextran in the system. Although these problems were anticipated to complicate the analysis of the PVP transport in dextran solutions, it appears that the various methods for evaluating diffusion coefficients from the Schlieren curves obey, to a certain degree, the criteria required for diffusional transport.

The formation of a boundary between the dextran solution and the dextran solution containing PVP 360 (concentration 5 kg m^{-3}) yields an apparently normal Gaussian distribution of the material detected by Schlieren optics. The various apparent diffusion coefficients obtained by an analysis of the Schlieren curves, which include diffusion coefficients obtained by the reduced height-area ratio method, the reduced second-moment and the width-at-half-height method, show the same qualitative behavior although quantitative differences do exist. This is seen in Fig. 7 where the

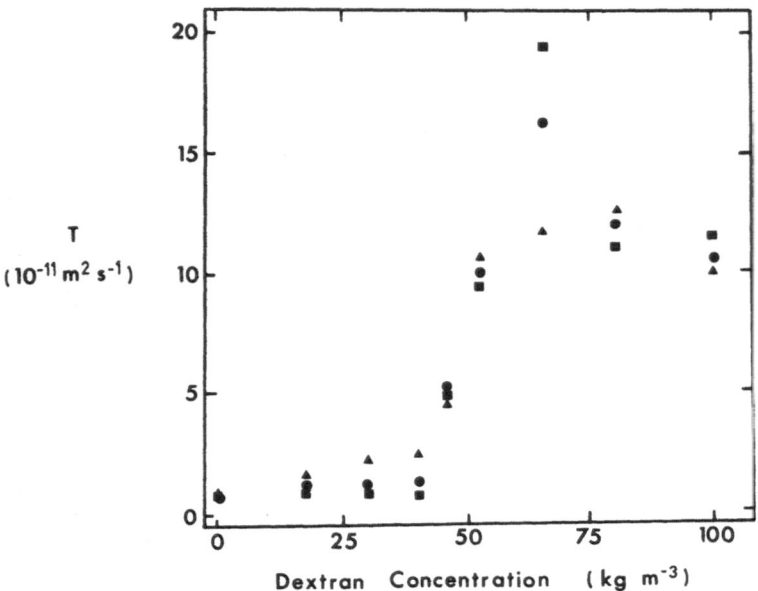

Fig. 7. Analysis of Schlieren curves for the transport of PVP 360 (initially at 5 kg m^{-3} to zero concentration gradient across the boundary) in dextran T10 solution. The transport coefficient T was obtained by diffusional analysis using the reduced height-area ratio method (●), the reduced second-moment method (▲) and the width at half-height method (■) [47]

measured coefficients are shown as a function of the dextran concentration (the concentration dependence of the PVP transport on the dextran concentration will be discussed below). Certainly, in none of the experiments reported in Fig. 7 has it been observed that the functions of the Schlieren patterns noticeably deviate from linear functions varying with time, thus indicating paradoxically that a diffusion-type process is operating in this system. Furthermore, there was no measurable displacement of the initial boundary during the course of the experiments. These apparently normal features of diffusion in this system have yet to be explained in the light of the evidence cited above as to the kinetics of the process.

However, one anomalous feature of this technique is that there occurs an apparent rapid removal of material from the concentration gradient at the boundary, as evidenced by a reduction in the area under the Schlieren curve. For the standard PVP/dextran T10 system, we observed a reduction of 20% during the 10 min after the formation of the initial boundary; no further changes in the area occurred after this initial event. This reduction in area is not accompanied by the appearance of refractive index gradients elsewhere in the cell [47]. The redistribution of material within the cell has been shown to occur by monitoring PVP 360 directly using absorption optics at 237 nm.

Absorption Optics

The time-dependent development of the initial absorption scans of the PVP transport in dextran, monitored at 237 nm, is shown in Fig. 8. The anomalous feature of these scans is that material which absorbs at 237 nm rapidly accumulates on the left-hand side of the boundary. This material appears to be evenly distributed in this region and would therefore not be detected by Schlieren optics. We have shown that accumulation of absorbing material on the LHS of the boundary is exactly balanced by the depletion of absorbing material on the RHS of the boundary.

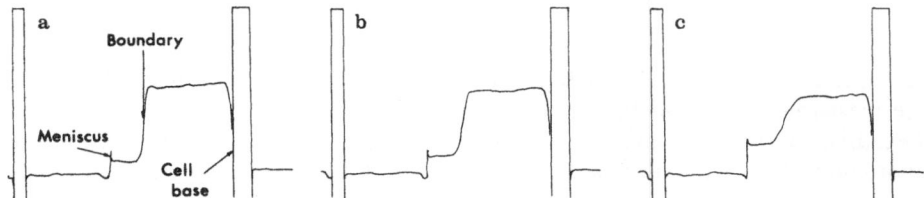

Fig. 8a–c. Time-dependent development of UV-scans, monitored at 237 nm in the analytical ultracentrifuge, of a dextran T 10 solution (concentration 135.2 kg m^{-3}) layered onto a solution of dextran T 10 (concentration 135.2 kg m^{-3}) and PVP 360, **a)** 180 s, **b)** 306 s, and **c)** 780 s after formation of the boundary layer [47]

An analysis of the increase in the absorbance on the LHS of the boundary, similar to the analysis of transport in the Sundelöf cell as shown in Fig. 5, indicates that the initial change is linear with time and falls off at later times (Fig. 9). For the experiment depicted in Fig. 9, which was carried out at a rotor speed of 4000 rpm, the initial PVP 360 transport rate was 3.8×10^{-3} m h^{-1}; this value is somewhat higher than the rates measured at unit gravity in the Sundelöf cell (Fig. 5) which gave a value of 1.4×10^{-3} m h^{-1}.

An unexpected feature of these systems is that whilst an initially rapid but partial redistribution of material appears to occur in the cell, the relaxation of the concentration gradient of absorbing material still remaining at the initial boundary appears follows a normal diffusion process. It should be stressed that the perceived concentration gradient at the boundary is a space-averaged parameter which does not reflect any changes in concentration that occur in the plane at right

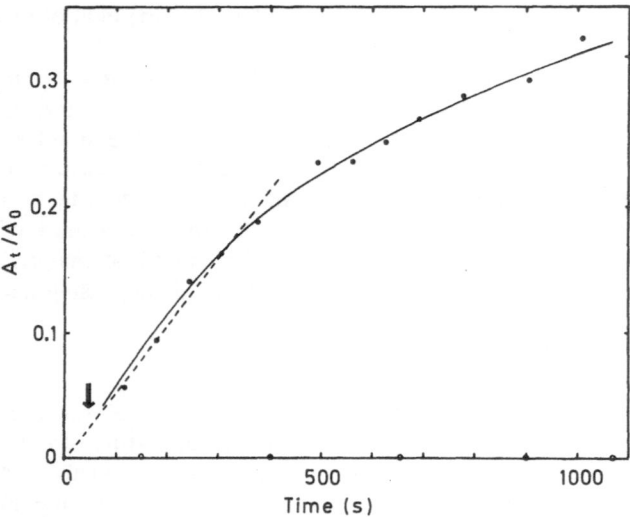

Fig. 9. Variation of A_t/A_0 with time. A_t is the average absorbance at 237 nm on the left-hand side of the boundary at time t and A_0 the initial absorbance at 237 nm on the right-hand side of the boundary for the solutions described in legend to Fig. 5 (●). The arrow indicates the time at which the rotor has its nominated speed. The corresponding experiment in which the dextran T10 concentration was lowered to 40 kg m^{-3} is also shown (○) [47]

angles to the optical axis. The lateral movement of this boundary is negligible within the measuring period so that the sedimentation of PVP is negligible. Diffusional spreading of the boundary is also minimal within the period of time (which is approximately 5–6 minutes) used for PVP transport rate measurements. An approximate evaluation of the diffusion coefficients made on a standard PVP 360/dextran T10 system gives a high value in the range of $100 \pm 2 \times 10^{-11}$ m^2 s^{-1}.

The studies concerned with the various analyses using the Schlieren and absorption optics are similar although they involve detection of different aspects of a multistep process. There are at least two modes of relaxation of the PVP concentration gradient a) one involving rapid redistribution of PVP in the cell which is only detected directly by absorption optics and b) a process which is essentially a local boundary phenomenon in which the observed concentration gradient gradually dissipates as in a diffusion-type process. The latter process is detected by both types of optics. Studies on the local distribution of material at the boundary as determined by a computer-colour indexing technique, which will be reported shortly, would indicate that these suggestions are correct [49].

3.3.2.4 Time-Dependent Development of [³H]PVP Distribution in a Vertical Test Tube

The spatio-temporal averaged transport measurements of [³H]PVP and [¹⁴C]sorbitol have been performed in a vertical test tube whose contents were fractionated at various times (Fig. 10). The distribution of both PVP and sorbitol appears to be sigmoidal, with gradual spreading with time. We have previously claimed [46] that

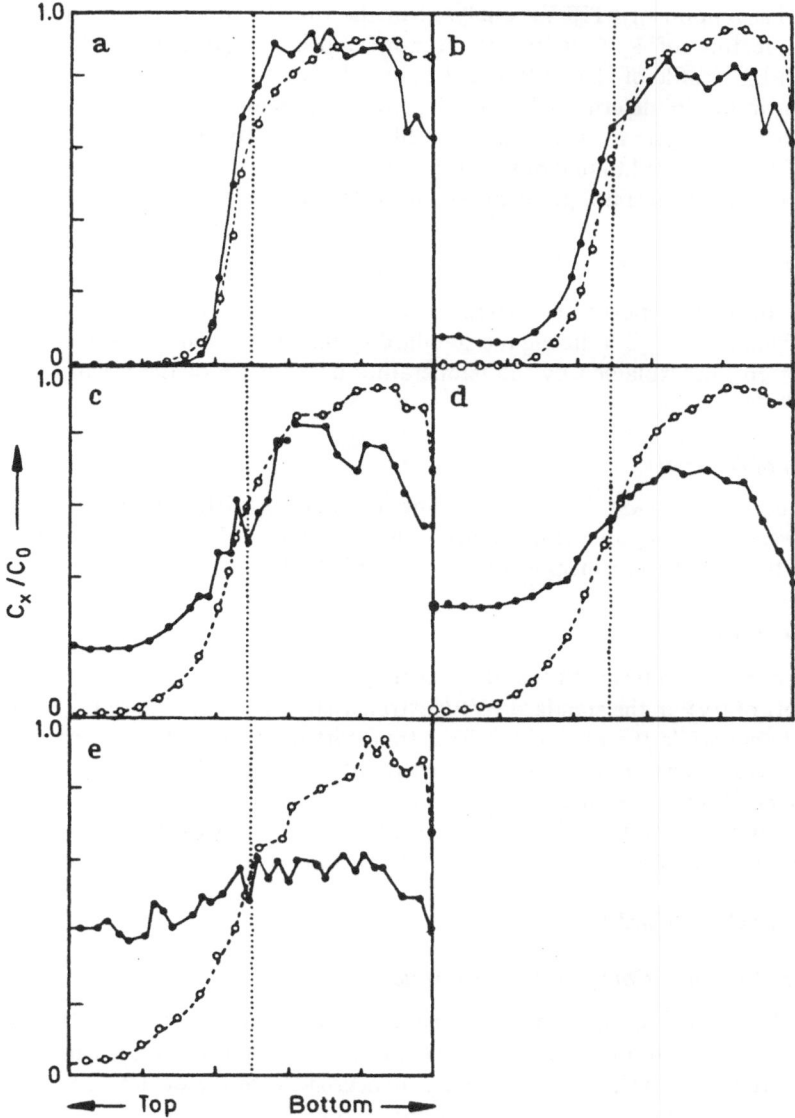

Fig. 10 a–e. Time distribution of [³H]PVP and [¹⁴C]sorbitol for solutions described in Fig. 5 with the inclusion of trace amounts of [¹⁴C]sorbitol in the bottom solution. The boundary was formed in plastic tubes (10 × 60 mm) between solutions of 19 mm in column height. A dense solution of tetra-bromoethane was placed into the bottom part of the tube. The contents of the tube were fractionated at various times by displacement with tetrabromoethane and the distribution of [³H]PVP (●) and [¹⁴C]sorbitol (○) was determined. The position of the original boundary is indicated by a dotted line. The ordinate represents the proportion of the radioactivity (in percent) in the sample, C_x, as compared to that originally present below the initial boundary C_0. The abscissa represents the ratio of the distance along the tube to the total length of the tube which was approx. 38 mm. The time of the fractionation was **a)** 11 min, **b)** 73 min, **c)** 134 min, **d)** 251 min, **e)** 490 min [47]

a wave-like distribution of [³H]PVP analysed in the tube-fractionated system is a characteristic feature of PVP transport; it now appears a good deal of noise observed with the profile of [³H]PVP in this system is due to an approximate 10% counting error in the ³H-determination. Perhaps the most remarkable feature of the PVP/dextran/sorbitol system, which is exemplified in Fig. 10, is the preferential spreading of PVP compared to that of sorbitol. After 8 h the PVP gradient has relaxed completely whereas the sorbitol gradient has not markedly changed [47].

3.3.2.5 Effect of Temperature

The variation of the transport rate of [³H]PVP with temperature (temperature range studied was from 5 to 35 °C) utilizing the standard system in the Sundelöf cell could, to a major extent, be explained by the temperature dependence of the viscosity of the solvent [50].

3.3.2.6 Effect of Geometry

Limited studies would suggest that, by decreasing the diameter of the initial boundary to 1 mm or less, the rate of polymer transport decreases. On the other hand, increasing the length of the cylindrical transport cell had no effect [51].

3.3.2.7 Effect of Gravity

The use of absorption optics with the ultracentrifuge has allowed us to monitor the rapid transport of PVP in the standard PVP/dextran system as a function of g. It was demonstrated that while the rate of the PVP transport increases with increasing g acting on the system, the rate is rather insensitive to the magnitude of the gravitational force. We found [51] that the linear time rate of the transport varies as $g^{0.19}$. Note, however, that although rapid PVP transport has been found at various values of g, we cannot be sure whether structured flows exist.

3.3.3 Variations of Standard Systems

3.3.3.1 Effect of Dextran Concentration and Molecular Weight

It has already been noted in Fig. 7 that the apparent diffusion coefficients of PVP 360 in dextran T10 increase abruptly after a critical concentration of dextran is reached.

We have carried out a wide range of studies concerned with the dextran concentration dependence of the transport of the linear flexible polymers and have varied both molecular mass and chemical composition of this component. Moreover, we have studied the effect of the variation of the molar mass of the dextran on the transport of the flexible polymers [51]. In general, the transport of these polymers in dextran solutions may be described on common ground. At low dextran concentrations the transport coefficients of the polymers are close to their values in the absence of the dextran and may even exhibit a lower value. This concentration range has been discussed in terms of normal time-independent diffusional processes in which frictional interactions predominate. We have been able to identify critical dextran concentrations associated with the onset of rapid transport of the flexible polymers. These critical concentrations, defined as C*, are summarized in Table 1. They are

Table 1. Critical dextran concentrations (kg m^{-3}) determined from the onset of rapid transport of flexible polymers (from ref. [51]).

Dextran type	T10	T20	T500
Flexible polymer			
PVP10	50–55	—	—
PVP40	45–50	25	5
PVP160	45	25	—
PVP360	45–50	25	5
PEG20	30	10	—
PVP14	50	—	8
Theoretical estimate[a]	118	71	35

[a] The theoretical estimate of C* is based on a closely packed array of equivalent spheres.

the same for different molecular weight PVP fractions in dextran T10 of T20 or T500. There is also a close agreement between the values of C* evaluated by the different transport methods [51]. Similar values of C* have been found for poly(vinyl alcohol) ($\bar{M}_w \sim 14,000$) (PVA14) as compared to PVP, but different critical concentrations are observed with poly(ethylene glycol) ($\bar{M}_w \sim 20,000$) (PEG20). We have also shown that C* depends on the molar mass of the dextran. Table 1 shows that the experimentally determined critical concentration of dextran required for the onset of a marked concentration-dependent enhanced transport of all flexible polymers studied is below the range of the critical concentrations calculated for closely packed encompassed spheres. The discrepancy between experimental and theoretical critical concentrations may depend on the intrinsic differences of effective dynamic interactions, the type of network formed and the particular average parameters used to calculate critical concentration [19]. It could be argued that the experimental value indicates a lower degree of interaction and overlap than required by the above model. Additionally, some of the dextrans studied are polydisperse and may therefore yield complex-averaged C* values.

In accordance with theoretical predictions of the dynamic properties of networks, the critical concentration of dextran appears to be independent of the molecular weight of the flexible polymeric 'diffusant' although some differences are noted when the behaviour of the flexible polymers used is compared e.g. the critical dextran concentrations are lower for PEG than for PVP and PVA. For ternary polymer systems, as studied here, the requirement of a critical concentration that corresponds to the molecular dimensions of the dextran matrix is an experimental feature which appears to be critical for the onset of rapid polymer transport. It is noteworthy that an unambiguous experimental identification of a critical concentration associated with the transport of these types of polymers in solution in relation to the onset of polymer network formation has not been reported so far. Indeed, our studies on the diffusion of dextran in binary (polymer/solvent) systems demonstrated that both its mutual and intradiffusion coefficients vary continuously with increasing concentration [12].

As the dextran concentration is further increased in the ternary system, the transport coefficient values often pass through a maximum, a second transitional point, and then gradually decrease. With increasing molecular weight of the dextran the concentration functions for the transport coefficient are displaced towards lower dextran concentration. This second transition point has also been previously discussed in terms of a dense hexagonal close packing of the encompassed volume spheres of dextran molecules [45]. The reasons for the strong increase in the transport coefficient and its gradual decrease at relatively high dextran concentrations are not clear. We have noted that in this dextran concentration regime the viscosity of the dextran solution is an important factor. We have found that when the PVP transport rates are plotted against the relative viscosity of the dextran solution, the variation of the molecular weight is effectively normalised, certainly up to concentrations leading to maximum transport rates (Fig. 11). This demonstrates that the rate of movement of PVP is directly related to the macroscopic viscosity of the solution. As the viscosity is increased the transport rate increases, reaches a maximum value and then decreases at high macroscopic viscosities [51].

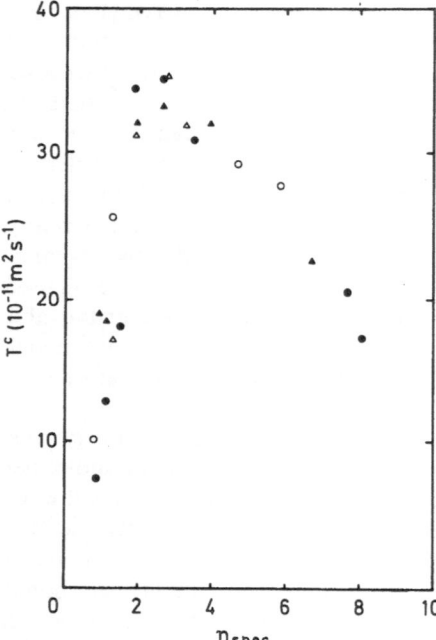

Fig. 11. Variation of T^c of $[^3H]$PVP360 transport in solutions of dextran of varying molecular weight as a function of the specific viscosity of the dextran solution (η_{spec}). Dextran $\bar{M}_w \simeq 1.04 \times 10^4$ (●); $\bar{M}_w \simeq 2.04 \times 10^4$ (○); $\bar{M}_w \simeq 6.94 \times 10^4$ (△); $\bar{M}_w \simeq 15.4 \times 10^4$ (▲) [51]

We emphasise again, however, that not too much significance should be placed on the absolute values of the transport coefficient obtained from these various experiments, due to the arbitrary nature of their evaluation. Nevertheless, these values clearly demonstrate the different trends and discontinuities that arise by varying the dextran concentration.

3.3.3.2 Density Gradient Stabilization

Variation of the PVP concentration gradient

When the PVP concentration in the lower solution is varied in the standard system, the [^{3}H]PVP transport rate increases with rising concentration up to 5 kg m^{-3} after which it approximately remains constant. The transport of [^{3}H]PVP even at low PVP concentrations (the lowest concentration used was 0.3 kg m^{-3}) is faster than in the absence of dextran [50].

Variation of the dextran concentration gradient

We have further stabilized the PVP/dextran system by increasing the concentration of dextran in the bottom layer, i.e. imposition of a dextran concentration gradient. The dependence of the [^{3}H]PVP transport on the magnitude of the concentration gradient, while maintaining the mean concentration of dextran across the boundary constant, is illustrated in Fig. 12. The corresponding difference in the initial macroscopic density ϱ_{init} across the boundary is also shown. The maximum rate of the PVP 360 transport is obtained near or at $\Delta\varrho_{init} = 0$. As the dextran concentration gradient is increased the transport rate of PVP decreases, but is still faster with a 20 kg m^{-3} dextran gradient than in the absence of dextran [51].

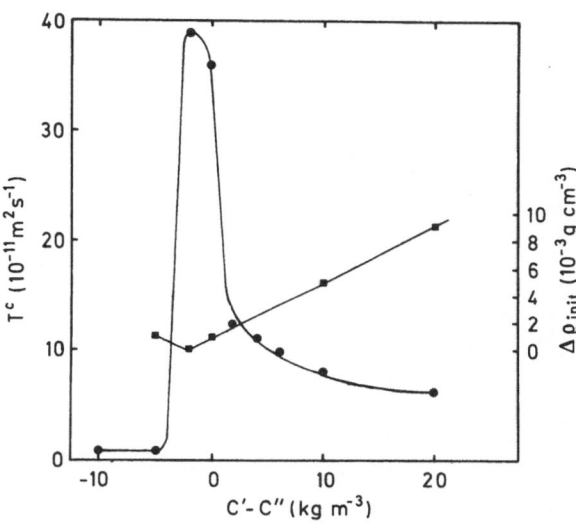

Fig. 12. Transport coefficient (●) for [^{3}H]PVP360 as a function of the initial concentration difference of dextran T10 that exists across the boundary, C'–C", where C' is the dextran concentration outside the capillary and C" the dextran concentration inside the capillary, the mean concentration of the dextran being 135.0 kg m^{-3}. The corresponding initial macroscopic density difference across the boundary $\Delta\varrho_{init}$ (■) is shown on the left-hand ordinate [51]

Inclusion of Low Molecular Weight Materials

The inclusion of low molecular weight solutes in the lower compartment of the standard system to supplement density stabilization does generally not affect the rapid transport of [^{3}H]PVP. However, this has not been studied systematically. When using 2 mol dm^{-3} NaCl or 10 kg m^{-3} sorbitol in the lower solution, no effect on polymer transport is observed [50,52]. Similarly, the use of 100% D$_2$O in the lower solution does not effect transport [46,50].

4 Identification of Structured Flows Associated with Rapid Polymer Transport

4.1 Morphology

Direct visualization of the fluxes involved was achieved using a blue-labelled PVP obtained by coupling PVP to remazol brilliant blue [46]. When a free boundary was formed in a Tiselius diffusion cell, a striking series of events was seen. After an

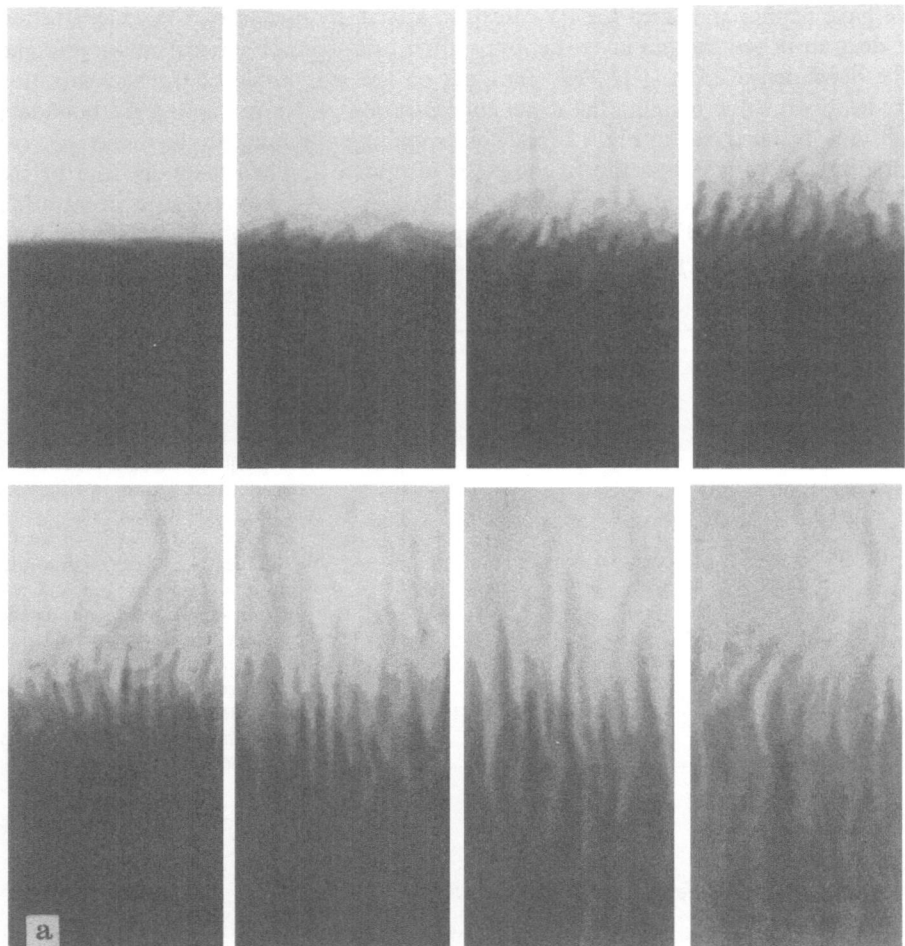

Fig. 13a. Time-dependent development of the macroscopic distribution of blue-labelled PVP at the boundary between solution of the standard system described in the legend of Fig. 5. The solution height above and below the boundary was approx. 1 cm, which is nearly the same as in the Sundelöf cell. The photographs were taken from left to right, top to bottom 0.08, 0.17, 0.32, 0.5, 1.0, 2.0, 3.0 and 4.0 h after formation of the boundary layer [52]

initial period, a number of uniformly spaced blue vertical fingers or spikes appeared at the boundary. These visible structures grew rapidly upwards and were associated with a concomitant development and propagation downwards of colourless fingers (Fig. 13a). Due to the heterogeneity of these structured flows no quantitative relationships between their movement and molecular transport, say of [³H]PVP, could be established.

On the other hand, with the inclusion of a low molecular weight stabilizer in the lower compartment, i.e. 2 mol dm⁻³ NaCl, the evolution of the structures or 'structured flows' was much more uniform and regular (Fig. 13b). The structures may be stable for several days.

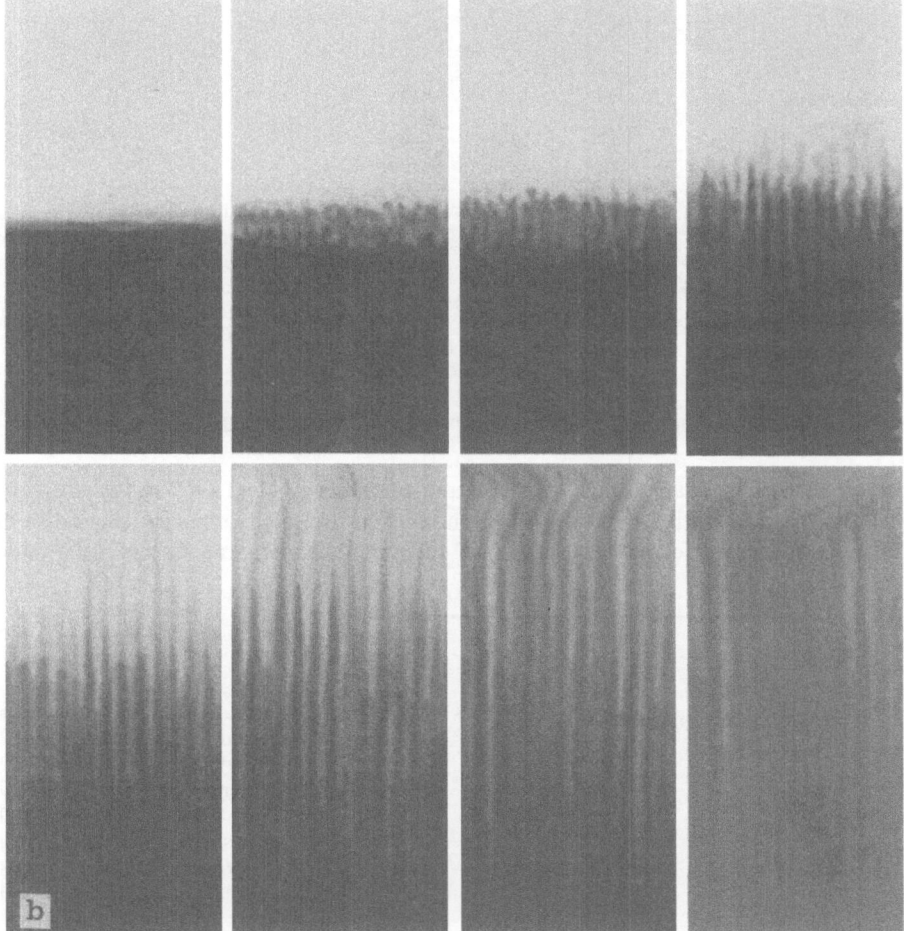

Fig. 13b. Time-dependent development of the macroscopic distribution of blue-labelled PVP at the boundary between 135 kg m⁻³ dextran (top) and 5 kg m⁻³ blue PVP, 135 kg m⁻³ dextran and 2 mol dm⁻³ NaCl (bottom). The solution height above and below the boundary was approx. 1 cm, which is nearly the same as in the Sundelöf cell. The photographs were taken from left to right, top to bottom 0.08, 0.17, 0.25, 0.5, 1.0, 2.0, 3.0 and 4.0 h after formation of the boundary layer [52]

To observe the formation of these structures from above, the experiments were performed in a petri dish with a glass coverslip. In this case, a regular mosaic pattern was observed within 10–15 min of the boundary formation (Fig. 14).

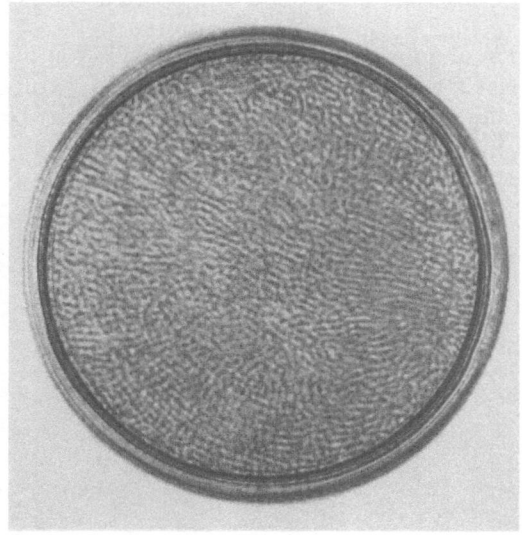

Fig. 14. Formation of structures in a petri dish ($\sim$2 cm in diameter) as seen from above. A boundary layer was formed between solutions described in legend of Fig. 13. The photograph was taken 15 min after formation of the boundary layer

4.2 Comparison between PVP Transport and Structured Flow Transport

The kinetics of the transport of [^{3}H]PVP and blue-dye labelled PVP in the standard system with and without NaCl is shown in Fig. 15. Not only are the rates of transport of the two forms of labelled PVP identical but their transport is not altered

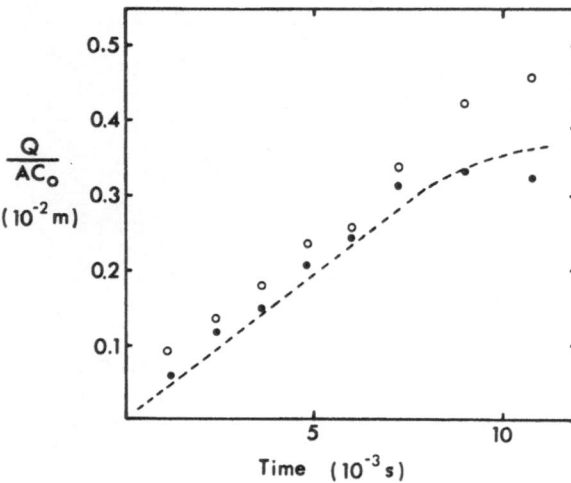

Fig. 15. Transport of various trace components including [^{3}H]-PVP ($\bullet$) and blue-labelled PVP ($\bigcirc$) over a boundary formed by layering 135 kg m^{-3} dextran T10 over a solution containing 5 kg m^{-3} PVP360, 135 kg m^{-3} dextran T10, 2 mol dm^{-3} NaCl, and the trace component whose transport is to be measured. For comparison, the transport of [^{3}H]PVP in the same system but without 2 mol dm^{-3} NaCl is shown (– – –)

by the presence of the low molecular weight stabilizer NaCl. The average rate of the [^{3}H]PVP transport in these systems is 1.4×10^{-3} m h^{-1}. Yet, the rate at which the structures move as measured by the movement of their heads is $\sim 10^{-2}$ m h^{-1}. The disparity between the rates of the movement of observable macroscopic structures and space-average polymer transport reveals that the most rapid frontal structures do not represent significant quantities of the transported blue-labelled PVP. We suggest that various forms of transport of the PVP occur, e.g. in the form of structured flows that move at different rates not only as separate structures but as structures within structures. It is probable that we only observe the most rapid phase of movement of these structures while most of the material is still associated with local initial boundary regions. This would be consistent with the bimodal nature of the relaxation of the PVP concentration gradient seen by absorption optics in the ultracentrifuge (Sect. 3.3.2.3.). Furthermore, measurement of the local distribution of blue-labelled material within each structure, as determined by a computer-colour indexing technique, would also indicate that these suggestions are correct [49].

4.3 When Do Structured Flows not Occur?

We have now identified the presence of structured flows in a wide variety of ternary systems of polymer/polymer/solvent [52,53]. In all cases associated with structured flow formation there was concomitant rapid transport of the polymer as compared to its behaviour in water. Indeed, even in the presence of dextran concentration gradients structures are formed which move relatively slowly but are nevertheless highly regular. The only conditions where structures have not been observed is at dextran concentrations below C* values where, incidentally, polymer transport is not rapid. (See also the low rate of transport of PVP 360 in a dextran T10 medium with a concentration of 40 kg m^{-3} as measured in the ultracentrifuge Fig. 9.) These studies confirm the striking correlation between this parameter and the onset of rapid polymer transport and structured flow formation.

5 Mechanistic Interpretations

5.1 General Considerations

The formation of structured flows has not been previously identified in polymer solutions. However, such a phenomenon seems to be acceptably equated with the general considerations of a 'dissipative structure' formation [54] and more recently classified under the title of 'synergetics' [55]. Nicolis and Prigogine [54] have described the general conditions required for the formation of ordered coherent structures. The basic general criterion necessary for dissipative structure formation is first of all that the system be an open system. Secondly, the system must be maintained 'far from equilibrium' (the appearance of such structures at equilibrium would be negligible). Thirdly, there are certain types of 'nonlinear' mechanisms operating between the various elements of the system. When these conditions are met, "certain types of fluctuations can be amplified and drive the system to a new regime different from the initial reference state." This regime is characterised by the appearance of a dissipative structure although the type of structure seen has not been predicted. It is of interest to compare our experimental findings with these criteria.

A good deal of similarity may be found in the morphological characteristics of the often quoted example of dissipative structure in fluid dynamics, that of the Raleigh-Bénard instability, and in the structured flows we observe in polymer systems. In the former, a horizontal fluid layer is heated from below. A temperature gradient is created which opposes the effects due to gravitational force. A critical temperature will be reached where density inversion occurs. This results in a systemic ordered convection as manifested by the appearance of well-defined cell-like structures as seen from above [56]. These structures are similar to those described in Fig. 14. The concepts of ordered convection arising from diffusion-mediated density inversion may also be applied to ternary systems containing polymers.

This will be elaborated in detail in the following section. However, it is of interest that the existence of concentration-dependent (implying a far-from-equilibrium condition) cross-diffusion terms creates a non-linear mechanism between 'elements' of the system, i.e. the flux of one polymer depends not only on its own concentration gradient but also on that of the other polymer component. This is consistent with two of the criteria required for dissipative structure formation. Furthermore, once a density inversion is initiated, by diffusion, it will be acted upon by gravity (as the system is 'open') to produce a structured flow. The continued growth, stability and maintenance of the structures once formed may depend on the lateral diffusion processes between neighbouring structures.

5.2 Polymer/Polymer/Solvent Systems

5.2.1 Initial Events Leading to Density Inversion

The possibility of an instability caused by external perturbations of the system (apart from gravity) such as mechanical vibration, light sensitivity and thermal effects has been effectively eliminated. The small increments in the concentration of one of the polymer components required for rapid transport would also make it unlikely that heats of mixing occurring during the initial transport creates adverse local temperature gradients in the system. We have also demonstrated that the partial specific volume of dextran is constant over a wide range of concentrations [8]. Thus, it is unlikely that diffusional transport leads to volume changes in the system.

A number of studies, associated with the diffusion of low molecular weight solutes in ternary systems, have considered the conditions for density inversion due to cross-diffusion effects [57, 58]. What was not recognised in these studies was the possibility of ordered convection occurring after density inversion had happened. However, the manifestation of the diffusion-mediated density instability (due to cross-diffusion effects between low molecular weight solutes) into dissipative structures was observed in the centrifuge by Vitagliano et al. [59]. (The question of 'interfacial tension' effects being the cause of an hydrodynamic instability at the boundary is an ambiguous one because the term has a broad meaning. It seems, however, that it may be equivalent, to a diffusion-mediated instability, (see footnote 5).

We have used similar principles to explain the occurrence of a density inversion arising from unperturbed diffusion without fluid motion in the case of ternary systems containing polymers. We have evaluated quantitatively all the ternary diffusion coefficients described in Eqs. (21) and (22) for the diffusion of 5 kg m^{-3} PVP 360 in dextran T500 ($M_w \simeq 500,000$) solutions of varying concentration up to 51 kg m^{-3}. These diffusion coefficients were calculated from Eqs. (23) to (26) from experimental measurements of μ_{ii} and μ_{ij} and are shown in Fig. 16. The striking feature of these diffusion coefficients is the marked concentration dependence of D_{23} at low dextran concentrations. Above a dextran concentration of 5 kg m^{-3}, D_{23} is greater than the principal diffusion coefficient D_{22}. On the other hand, D_{33} is consistently about twice as large as D_{32} over the dextran concentration range studied. The effect of a

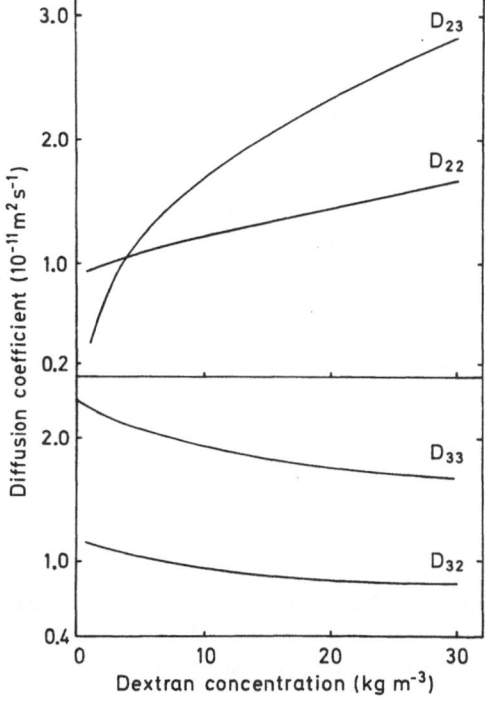

Fig. 16. Graphically smoothed data for calculated ternary diffusion coefficients in a system with a uniform concentration of dextran T500 ($\bar{M}_w \sim 500,000$) with an imposed 5 kg m^{-3} concentration gradient of PVP 360; the dextran concentration is varied [36]

5 Interfacial turbulence causing hydrodynamic instability between two solutions has long been known as the Marangoni effect. It is thought that this effect is induced by longitudinal variation of interfacial tension, normally occurring between two immiscible or slightly immiscible fluids [60]. However, it is difficult to distinguish molecular behaviour giving rise to interfacial tension and its variation to molecular transport between two miscible fluids, i.e. random molecular fluctuations at the interface. Owing to the broad definition of interfacial tension it is probable that the Marangoni effect and diffusion-related hydrodynamic instabilities occurring in the case of miscible solutions have much in common and cannot be easily distinguished.

non-zero value of D_{23} can be immediately visualised as illustrated in Fig. 17a. In this case the movement of PVP induces an inverted concentration gradient of dextran which, in turn, may give rise to a density inversion. It is the magnitude of D_{23} which primarily determines whether inversion occurs. As seen in Eq. (24), D_{23} is a function both of thermodynamic parameters, namely μ_{kh}, and phenomenological terms. Evaluation of these individual parameters has clearly indicated that the dominant term in Eq. (2) is c L_{22} [31]. Furthermore, we have shown that the numerical value of f_{32} in Eq. (31) is relatively low so that $L_{22} \simeq m_2/f_{21}$, which is approximately the same phenomenological coefficient describing the binary diffusion of dextran, i.e. in the absence of PVP. It is therefore the c-term which primarily determines the magnitude of D_{23}. At the concentrations used, the c-term is dominated by the thermodynamic interaction parameter μ_{32}. This parameter characterizes the excluded-volume interactions between dextran and PVP.

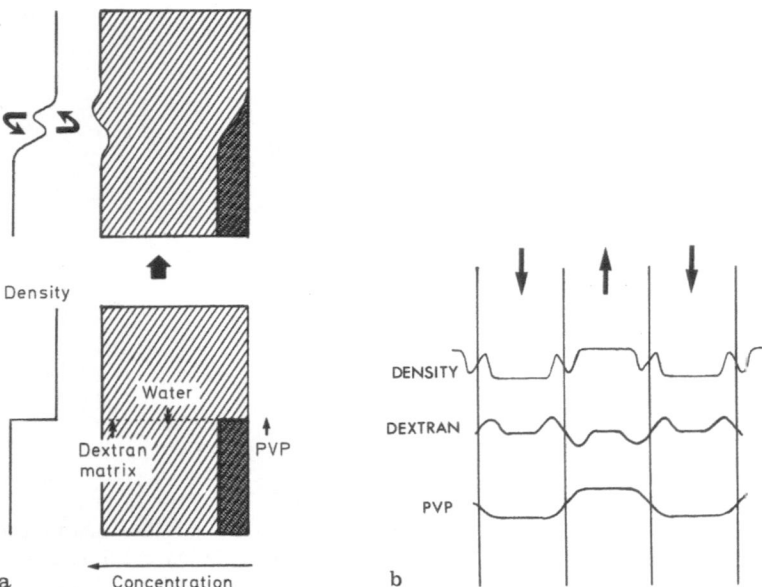

Fig. 17a. Schematic diagram of the concentration and density gradients occurring in the system described in the legend of Fig. 16

Fig. 17b. Schematic diagram of possible concentration and density gradients in countercurrent structured flows (for details see text)

When all ternary diffusion coefficients are known, we can predict the concentration gradients of all components [35] and therefore density gradients in the system at any time.

The general conditions necessary to ensure gravitational stability at all times during the free diffusion process and at all positions in the diffusing boundary have been calculated by Wendt [57]. These conditions are considered as complex functions, i.e.

$$(1 - \bar{V}_3) K_3^- + (1 - \bar{V}_2) K_2^- \geq 0$$

and

$$(1 - \bar{V}_3) K_3^+ + (1 - \bar{V}_2) K_2^+ + \frac{\theta_-}{\theta_+} ((1 - \bar{V}_3) K_3^- + (1 - \bar{V}_2) K_2^-) > 0$$

where

$$K_i^\pm = \pm [(\theta_\pm - D_{ii})\Delta C_i - D_{ij}\Delta C_j]/2(\theta_+ - \theta_-) \qquad (i \neq j \neq 1),$$

$$\theta_\pm = \frac{1}{2} \{D_{22} + D_{33} \pm [(D_{33} - D_{22})^2 + 4D_{32}D_{23}]^{1/2}\}$$

and $\Delta C_i = (C_i)_B - (C_i)_A$ where the subscripts A and B denote solutions below (B) the boundary and above (A) the boundary. Computation of density gradients and therefore graviational stability by means of these equations utilizing the ternary diffusion coefficient data given in Fig. 16 reveals that inversion occurs for the PVP 360/dextran T 500 ternary system at dextran concentrations above 5.0 kg m^{-3}. At lower dextran concentrations, density inversion does not occur. This critical dextran concentration characterizing the transition of a gravitationally stable system to an unstable one is almost identical to the critical dextran concentration determined for the onset of rapid transport as shown in Table 1. The reason for this correlation is not clear. However, the correspondence between the onset of rapid transport and the density inversion is strong evidence that these two processes are linked.

Another approach employed to establish the occurrence of a density inversion between the two solutions subsequent to boundary formation involves dialysis between the two solutions [50]. The dialysis membrane is impermeable to the polymer solutes but permeable to the micromolecular solvent, H_2O. Transfer of water across the membrane occurs until osmotic equilibrium involving equalization of water activity across the membrane is attained. Solutions equilibrated by dialysis would only undergo macroscopic density inversion at dextran concentrations above the critical concentration required for the rapid transport of PVP 360 [50]. The major difference between this type of experiment and that performed in free diffusion is that in the former only the effect of the specific solvent transport is seen which is equivalent to a density inversion occurring with respect to a membrane-fixed or solute-fixed frame of reference. Such restrictions are not imposed on free diffusion where equilibration involves transport of all components in a volume-fixed frame of reference. The solvent flow is governed specifically by the flow of the polymer solutes as described by Eq. (3) which, on rearrangement, gives

$$J_1 = -\frac{1}{V_1} (J_2 V_2 + J_3 V_3)$$

In defining

$$J_1 = -D_1 \frac{\partial m}{\partial x}$$

we can easily obtain the diffusion coefficient of water D_1 as

$$D_1 = \left(D_{33} + D_{23} \frac{V_2}{V_3} \right)$$

In the ternary system, therefore, the diffusional flux of water is determined by two of the ternary diffusional coefficients. For a binary system, it was shown earlier that the mutual diffusion of solvent and solute is identical and essentially independent of the magnitude of the osmotic pressure gradient across the boundary [30].

5.2.2 Structured Flow Formation

Far less information is available on the mechanism of the growth and stabilisation of structured flows once the nucleation process of coupled diffusion-mediated density inversion occurs. At this stage, only speculative arguments can be offered.

The mere fact that a density inversion occurs through coupled diffusion of polymer components is not a sufficient condition for the generation of instability associated with fluid motion. For example, the system may be stabilised by either fluctuations in the movements of the solute leading to dissipation of the density inversion or local viscosity gradients.

In simple terms, the onset of fluid motion may be qualitatively understood in terms of the sedimentation of an object in a fluid at unit gravity. The total force F acting on the object of critical volume V_c is equal to the sum of the buoyancy force and frictional force such that

$$F = b(V_c(D), \varrho(D), g) + f(V_c(D), \eta, v)$$
$$\text{(buoyancy)} \qquad \text{(frictional)}$$

where both the critical volume (V_c) and density (ϱ) are a function of the diffusion (D) or the residence time of the solute components in the volume element. The buoyancy force is opposed by a frictional force on the volume element moving at a velocity v through a medium of macroscopic viscosity η. We have already established that the macroscopic viscosity is of primary importance in regulating the rate of movement of the structures (Fig. 11) although the relationship is a complex one. Another important factor in the formation of the structure flow is the value of ϱ, determined by the relationship between the magnitude of the inverted density gradient (Fig. 17a) and the distance over which it operates.

If initial migration of critical volume elements occurs, the regularly spaced volumes of ascending and descending motions are generated through long-range co-operative interactions of pressure within the solution.

We speculate that continued growth and stabilisation of the structure occurs due to cross-diffusion effects across boundaries separating ascending and descending flows. The type of mechanism envisaged is exactly analogous to that described in Fig. 17a. The concentration and macroscopic density profiles across the boundary are indicated in Fig. 17b. The thin layers on either side of the boundary are now exposed to density instability brought about by lateral diffusion effects. The layers differ markedly in density with respect to one another and to the density within the interior of each structured flow. The thin layers move in opposite directions in a gravitational field and may facilitate their mutual motion. (In an analogous manner, it is known that when neutral buoyant particles are added to a sedimenting suspension, they quite markedly accelerate the sedimentation rate [61, 62].) Due to the surface instability, relatively rapidly moving outer sheaths then result which drag along the inner solution of different density. With time the outer sheath may plume off the main core of the structure giving rise to a turbulent effect with the initial structures formed in the system (designated as front structures). Presumably, if one sheath is removed, another forms on the outer surface of the structure which will continue to drive the structure vertically. The fluid within the structured flow will move at a varying velocity. A similar situation could be envisaged for the reverse of a viscosity experiment where a solution flows in a capillary (fixed) under the action of gravity. In the case of structured flow the outer sheath could be viewed as a moving capillary which causes the motion of fluid within it. This situation would lead to spectrum of flows within a structure migrating at different rates. This is quite consistent with our findings of the marked disparity between the rates of movement of the frontal structures and the spatiotemporal average rates of polymer movements within these structures. Finally, the reasons why the front structures have characteristic 'mushroom-like heads' are not clear although it is only the rapidly moving front structures which exhibit this morphological characteristic.

5.3 Polymer/Low Molecular Weight Solute (Monomer)/Solvent (System A)

We have shown above that for component configurations characteristic of system A, where solute densities are different, structured flows and rapid polymer transport may occur. These are more or less independent of the relative mobilities of the polymers involved.

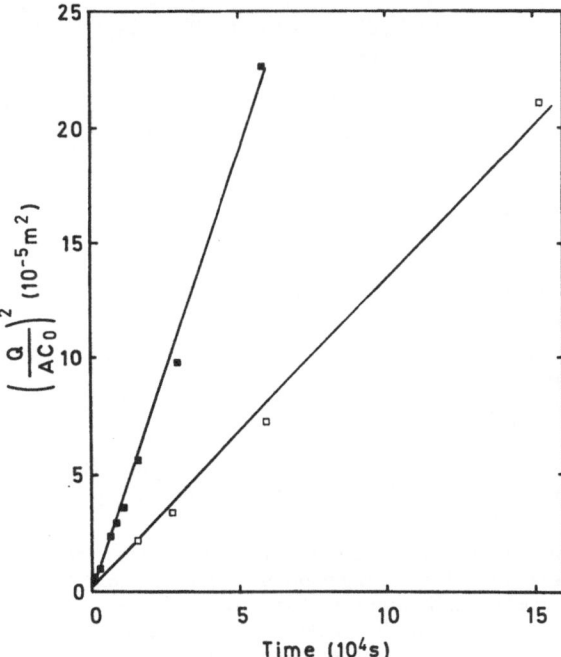

Fig. 18. Transport of [^{3}H]dextran over a boundary formed by layering 100 kg m^{-3} dextran T10 over a solution containing 100 kg m^{-3} dextran T10 (with [^{3}H]dextran T10) and 50 kg m^{-3} sorbitol (■–■). The same system as above was used except that sorbitol was replaced by 5 kg m^{-3} dextran T10 (□–□)

We have extended studies on system A-containing-polymer but with solute components of the same density and chemical composition varying only in their relative mobilities. We have demonstrated the rapid polymer transport in these systems. These studies have only been briefly reported [63]. The results obtained by measurements on a system containing uniform concentration of dextran T10 throughout (i.e. 100 kg m^{-3}) with an imposed gradient of sorbitol concentration at 50 kg m^{-3} demonstrate this effect. Fig. 18 describes the kinetics of the dextran transport has which not only exhibits a 'diffusion-like' behaviour but also occurs relatively rapidly, namely abbout twice the rate of the [^{3}H]dextran transport in intradiffusion. This rapid transport of dextran, induced by the sorbitol gradient, indicates a strongly coupled diffusion process. (Preliminary measurements suggest that the cross-diffusion coefficients between sorbitol and dextran are large). This is accompanied by a slow but highly ordered development of structured flows over a certain of time as shown in Fig. 19. The identification of rapid transport and structured flows in this type of

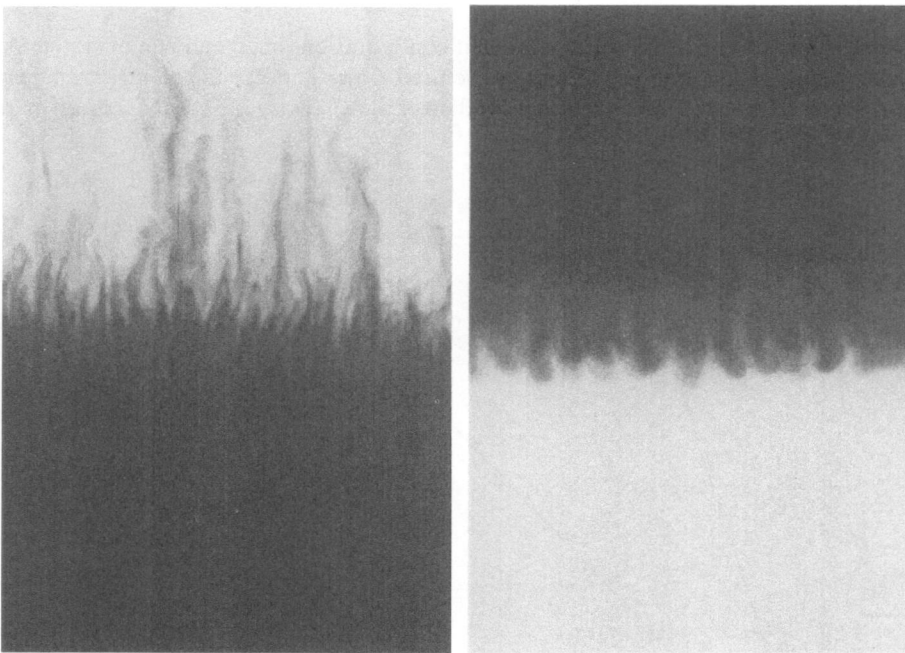

Fig. 19. Time-dependent development of structured flows in a dextran/sorbitol system (see legend of (Fig. 18) with blue-labelled dextran initially introduced into the bottom solution. The photograph was taken 2 h after formation of the boundary layer

Fig. 20. Structured flows developed between the bottom solution containing dextran with 6.146×10^{-2} g/g phosphate buffer at pH 6.0 and dextranase with 10 units/g phosphate buffer ($= 2.63 \times 10^{-5}$ g/g phosphate buffer) and the top solution containing dextran with 3.952×10^{-2} g/g phosphate buffer (including blue dextran). The boundary was formed 3 min after the dextranase was added to the dextran solution. The photo was taken 30 min after formation of the initial boundary [53]

system containing components of the same chemical composition points to the general nature of this phenomenon which may be particularly important in a biological system where the polymers exhibit a wide range of molecular weight heterogeneity.

A recent paper by McDougall and Turner [64] has offered a theoretical interpetation of structured flow formation in the dextran/sorbitol system. This is based on a phenomenological analysis of spontaneous growth of an instability that occurs at the boundary. Although the mechanism for the formation of the initial instability is undefined in their treatment (as it may occur from a wide range of sources including temperature fluctuations, mechanical vibrations etc.), the growth of the initial process and stability of the structured flows has been considered in terms of cross-diffusion effects. Structured flow formation in systems studied by us are singularly determined by the properties of the solutes used and not by extrinsic factors other than gravity.

A comparison of the data obtained by using McDougal and Turner's treatment of the prediction of structured flow formation in the PVP-dextran system with the ternary diffusion data in Fig. 16 indicates that their theory is not compatible with our observations [36].

5.4 Interdiffusion (System B)

The interdiffusion of low molecular weight solutes such as sugars and salts or that of heat and salt is known to give rise to structured flows which have some features in common with those discussed in connexion with system A. This type of motion and the solute configuration have been described by Turner [65] under the heading of so-called 'double-diffusive convection' which requires 1) the fluid must contain two (or more) components having different molecular diffusivities and 2) these components must make *opposing* contributions to the vertical density gradient'.

Studies of interdiffusion have been extended to polymer systems. Structured flow formation was detected for these systems in zonal centrifugation experiments [66,67]. If a polymer solution is layered on top of a more dense solution containing a low molecular weight solute, then interdiffusion of the salt into the polymer solution eventually gives rise to density inversion, and droplets of polymer solution will sediment. This effect has often been described as 'droplet sedimentation' or the 'turnover effect' [68].

The conditions required for ensuring the stability of flows as well as their properties in these interdiffusing systems have been widely discussed and are beyond the scope of this paper. The major difference between the development of flows in system A as compared to system B is the seeming requirement of coupled diffusion in system A; such a requirement is not necessary for flows generated by density instabilities in system B [68-70]. While a stability analysis based on density inversion through interdiffusion with no cross-diffusion imposes too limited conditions upon the solute distribution and boundary stability [57,68], a stability analysis performed on the basis of the phenomenological description of spontaneous growth of an undefined random fluctuation without cross-diffusion appears to introduce satisfactory limits to the stability of interdiffusing systems. [69,70] However, we have recently demonstrated that density inversion through cross diffusion effects in interdiffusing systems may also account for the structured flow formation of the latter [71]. Additionally, we have demonstrated [71] that the rates of structured flow formation in these systems is strongly influenced by the critical concentration of the polymer as found for system A.

6 Model Biological Systems

6.1 Biopolymers and Structured Flows

The simple mechanisms by which structured flows are initiated, e.g. through instabilities formed by cross-diffusion or interdiffusion, reveals their potential ubiquitous nature. We have employed a number of multicomponent systems containing biopolymers and identified structured flow formation and rapid transport [53]. These systems include protein/amino acid, polysaccharide/monosaccharide, cartilage proteoglycan/chondroitin sulfate and polysaccharide/degrading enzyme. The use of these systems has revealed the general nature of the phenomenon associated with multicomponent systems of the polymer/monomer and polymer/oligomer type.

It was of interest then to generate these multicomponent systems in situ. This can be performed in the presence of a polymer-degrading enzyme as shown in Fig. 20. In this experiment, there was initially a marked concentration gradient of dextran in the system which, as a binary system, is stable, and only diffusion of dextran and water occurs. However, when dextranase is added to the solution below the initial boundary, the whole system spontaneously generates structured flow. Dextranase is present at very low concentrations and negligibly contributes to the physicochemical parameters of the system. Furthermore, the tendency of the enzyme to move into the solution above the initial boundary by diffusion, is restricted by the relatively high dextran concentration. The enzyme causes degradation of the dextran (in the bottom layer) to maltose units. The resulting distribution of the polymer in the upper layer and oligomer in the lower one is unstable and ultimately generates structured flows in the system and causes an appreciable increase in the surface area of the boundary between the original upper and lower solutions.

6.2 Rapid Transport of Large Molecules

One of the striking features of a dynamic structured flow system is that it may act as a vehicle for the transport of material. We have added a range of different radiolabelled molecules of various size at trace concentrations to the solution below the initial boundary of the standard PVP dextran system described in Fig. 5 and measured their transport rates; these are compiled in Table 2. It is obvious that with increasing molecular weight and hence decreasing intrinsic diffusive mobility of the molecule its transport rate in a structured flow system becomes much greater. A limiting rate is reached corresponding to the size of the trace molecule used, i.e. all molecules with a molecular weight higher than that of PVP 360 have same rate of movement in the structured flow system.

Table 2. Transport of various trace solutes in a structured flow system formed by the presence of a 5 kg m^{-3} PVP 360 concentration gradient in a 135 kg m^{-3} dextran T10 solution (adapted from Ref. [50])

Solute	Molecular weight	Apparent diffusion coefficient in the presence of structured flow system	Enhanced diffusion transport rate[a]	Linear transport rate (10^{-8} m s^{-1})
HTO	20	174	1.1	
^{22}Na$^+$	22	98	1.2	
[^{14}C]Sorbitol	182	42	1.2	
[^{3}H]Raffinose	505	46	1.6	
[^{14}C]PEG	4000	49	5.0	
[^{3}H]Dextran	10,000	94	26.0	14
[^{125}I]Bovine Serum Albumin	67,000	210	124.0	26
[^{3}H]Dextran	150,000	365	~200	21
[^{125}I]Collagen	300,000	~1000	~1500	39
[^{3}H]PVP 360	360,000	460	~400	38
[^{3}H]Hyaluronate	4,000,000	~5000	~20,000	42

[a] Calculated from the ratio of the apparent diffusion coefficient in the structured flow system to the diffusion coefficient of the trace solute in free solution.

These results are schematically illustrated in Fig. 21. Substances of low mobility are retained in the ascending structure. Their vertical transport occurs rapidly and is basically determined by the bulk movement of the structures. On the other hand, highly mobile substances such as sorbitol undergo a rapid lateral diffusive exchange between ascending and descending structures and hence the net transport in the vertical direction is low and governed by lateral diffusion.

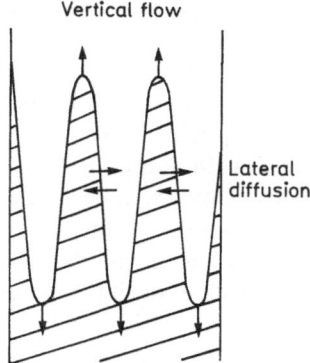

Fig. 21. Schematic representation of the molecular weight dependence of rapid transport involving structured flows as vehicles for the transport. Large molecules are retained in vertically migrating rapid flows and undergo only slight or no lateral diffusion. Small molecules, on the other hand, are not retained in rapidly moving vertical flows but diffuse laterally into a vertical flow migrating into the opposite direction

6.3 g-Dependence and Directionality of Flows

We have discussed in Section 3.3.2.7 that the linear rate of the PVP transport varies as g$^{0.19}$ in a structured flow system.

Experiments have not yet been performed at reduced gravity (g < 1). From the power relationships it may be predicted that the rate of structured flow transport is particularly insensitive to reduced g values. For example, even at 10^{-5} g, a value corresponding to centripetal forces in linear orbit [72], it is predicted that the rate will only be reduced to approximately 0.1 times its value at unit gravity. Indeed, convectional processes have been shown to occur at these weak gravitational fields. Alternatively, these reduced rates could be understood to correspond to very low angles with respect to the horizontal plane by resolving unit gravity into reduced values. For example, if a structured flow was mechanically restricted to move at an angle of 6×10^{-5} degrees from the horizontal plane, then the effective gravity acting in this direction would be 10^{-5} g. Thus, the structure would predictably move at approx. 0.1 times its value at unit gravity. A 10 fold reduction in the transport rate is not a significant reduction considering that spatio-temporal average transport measurements of the PVP transport in the Sundelöf cell indicate that the movement of PVP is 100 times faster than normal diffusion at unit gravity and that structured flow rates may be about 10000 times higher than for normal PVP diffusion. It is therefore conceivable that these structures, while driven by gravity, may still move at relatively high rates (as compared to normal diffusional migration) at angles close to the horizontal axis.

7 Future Work and Unsolved Problems

Rapid polymer transport and associated structured flow formation are multistep processes. These processes may include 1) initial diffusion of components across the boundary 2) inversion of density, 3) convective motions occurring in regions that are unstable with respect to density, 4) birth and nucleation of structured flows, 5) development of visible structured flows, and 6) movement and maintenance of structured flows over longer periods. A clear delineation of any of these steps has not been achieved so far. Future work will be concerned with the development of systems in which these individual steps may be studied in more detail.

A number of interesting experimental correlations have been established although their meaning is not yet clear. For example, we have found a strong interdependence between the critical concentration of the dextran network and the onset of rapid transport and structured flow formation. Additionally, a density inversion, being the initial stage of structured flow formation, may arise from cross-diffusion effects. In order to establish a molecular mechanism for these correlations a number of variables have to be considered, i.e. the conformation and flexibility of the interacting diffusing molecules, the rate at which they normally diffuse in free solution, and the nature of the interactions which dominate the cross-diffusion terms. Studies made so far suggest that the latter interactions are mainly entropic, based on excluded volume effects [36] (Sect. 5.2).

Our studies may contribute to the understanding of molecular organisations associated with the biological evolution and morphogenesis of macroscopic structures. These processes are only indirectly related to genetic information (which does not specify where molecular material is to be localised). Furthermore, these processes involving passive, non-metabolic interactions are often extracellular [2]. Gene-mediated processes seem to be restricted to the coding of proteins or enzymes which may be involved in the synthesis of polymers that interact extracellularly or participate in interactions themselves, leading to the formation of much larger aggregates. In this context, we may now study the phenomenon of how large polymers or structures are able to move rapidly or be relocated to their extracellular matrix, in relation to the cell.

It is clear that by developing and understanding the basic parameters of the system we will eventually delineate those biological systems in which this phenomenon may occur. For example, the transport in the nerve axon has been studied in detail [73] but it has not been explained how flow can occur in both directions and how particulate material is transported more rapidly than soluble proteins and low molecular weight material. These features are exactly the type of transport features that we have described in our polymer system. We have recently proposed a paradigm for axonal transport based on the essential features of structured flow of polymers [74]. Recent observations on the transport of high molecular weight material in certain membranes [41] have indicated a surprisingly high transport rate compared to the free diffusion of the studied compounds. It will be interesting to investigate whether the transport by structured flows can also occur in regular membranes.

The question whether density instabilities driven by gravity are of some importance to higher biological organisms has not been clarified. The ordered convective behaviour of microorganisms due to similar mechanisms has been well established [75]. It is conceivable that other forces such as hydrodynamic flow may cause similar structures to be formed. The question also arises whether such macroscopic structures or convectional processes are at all important in the organisation and transport on a microscopic dimensional level, e.g. in cytoplasmic flow. The flows are regulated by viscous drag between countercurrent flows and by interfacial effects of the solution and fixed boundaries of the system. It seems that in order to invoke the occurrence of a similar phenomenon on a more microscopic level, frictionless surfaces and reduced viscous drag are required.

Acknowledgements: This work was supported by the Australian Research Grants Committee. We acknowledge the excellent technical assistance of Gregory Checkley and Geoffrey Wilson. Much of this work has come from studies made in close collaboration with Professor T. C. Laurent and his colleagues at Uppsala University, Sweden. Unfortunately, due to heavy University commitments, he was unable to contribute as co-author to this paper.

References

1. Preston, B. N., Laurent, T. C., Comper, W. D. In: Molecular Biophysics of the Extracellular Matrix (eds.) Arnott, S., Rees, D. A., Clifton, New Jersey, Humana Press 1983
2. Comper, W. D., Laurent, T. C.: Physiol. Rev. *58*, 255 (1978)
3. Kirkwood, J. G. et al.: J. Chem. Phys. *33*, 1505 (1960)
4. Spiegler, K. S.: Trans. Faraday Soc. *54*, 1409 (1958)
5. Kedem, O., Katchalsky, A.: J. Gen. Physiol. *45*, 143 (1961)
6. Onsager, L.: Ann. N. Y. Acad. Sci. *46*, 241 (1945)

7. Bearman, R. J.: J. Phys. Chem. *73*, 186 (1961)
8. Comper, W. D., Van Damme, M.-P. I., Preston, B. N.: J. Chem. Soc. Faraday Trans. 1, *78*, 3369 (1982)
9. Larm, O., Lindberg, B., Svensson, S.: Carbohydr. Res. *20*, 39 (1971)
10. Granath, K.: J. Colloid. Sci. *13*, 308 (1958)
11. Basedow, A. M., Ebert, K. H.: J. Polym. Sci., Polym. Sym. *66*, 101 (1979)
12. Preston, B. N. et al.: J. Chem. Soc. Faraday Trans. 1, *78*, 1209 (1982)
13. Laurent, T. C. et al.: Eur. J. Biochem. *68*, 95 (1976)
14. Edmond, E. et al.: Biochem. J. *108*, 775 (1968)
15. Ogston, A. G., Woods, E. F.: Trans. Faraday Soc. *50*, 635 (1954)
16. De Gennes, P. G.: Macromolecules *9*, 587 (1976)
    De Gennes, P. G.: Macromolecules *9*, 594 (1976)
    De Gennes, P. G.: Nature *282*, 367 (1979)
17. Laurent, T. C., Ryan, M., Pietruszkiewicz, A.: Biochim. Biophys. Acta. *42*, 476 (1960)
18. Graessley, W.: Adv. Polym. Sci. *16* (1974)
19. Aharoni, S. M.: J. Macromol. Sci. Phys. B. *15*, 347 (1978)
20. Nyström, B., Roots, J.: J. Macromol. Sci., Rev. Macromol. Chem. *C19*, 35 (1980)
21. Maron, S. H., Nakijima, N., Krieger, I. M.: J. Polym. Sci. *37*, 1 (1959)
22. Ogston, A. G., Preston, B. N.: Biochem. J. *182*, 1 (1979)
23. Franks, F. et al.: Cryo-Lett. *1*, 104 (1979)
24. Arnott, S., Winter, W. T.: Fed. Proc. *36*, 73 (1977)
25. Adam, M., Delsanti, M., Pouyet, G.: J. Phys. Lett. *40*, L-435 (1979)
26. Wik, K.-O., Comper, W. D.: Biopolymers *21*, 583 (1982)
27. Valtasaari, L., Hellman, E.: Acta Chem. Scand. *8*, 1187 (1954)
28. Sundelöf, L.-O.: Ber. Bunsenges. Phys. Chem. *83*, 329 (1979)
29. Einstein, A.: Investigations on the theory of the Brownian Movement, Dover Publications, 1956
30. Van Damme, M.-P., Comper, W. D., Preston, B. N.: J. Chem. Soc. Faraday Trans. 1, *78*, 3357 (1982)
31. Tanaka, T., Fillmore, D.: J. Chem. Phys. *70*, 1214 (1979)
32. Munch, J. P. et al.: J. Phys. *38*, 971 (1977)
33. Candau, S., Bastide, J., Delsanti, M.: Adv. Polym. Sci. *44*, 27 (1982)
34. Miller, D. G.: J. Phys. Chem. *63*, 570 (1959)
35. See e.g. Cussler, E. L.: Multicomponent Diffusion, Amsterdam, Elsevier 1976
36. Comper, W. D. et al.: J. Phys. Chem. (in press)
37. Laurent, T. C. et al.: Biochim. Biophys. Acta *78*, 351 (1963)
38. Preston, B. N., Snowden, J. McK.: Proc. Royal Soc. Lond. *A333*, 311 (1973)
39. Ogston, A. G., Preston, B. N., Wells, J. D.: Proc. Royal Soc. London *A333*, 297 (1973)
40. Laurent, T. C. et al.: Eur. J. Biochem. *53*, 129 (1975)
41. Cumming, G. J., Handley, C. J., Preston, B. N.: Biochem. J. *181*, 257 (1979)
42. Cussler, E. L., Lightfoot, E. N.: J. Phys. Chem. *69*, 2875 (1965)
43. Preston, B. N., Snowden, J. McK., in: Biology of Fibroblast (eds.) Kulonen, E., Pikkarainen, J., p. 215, New York, Academic 1973
44. Kitchen, R. G., Preston, B. N.: Ph. D. Thesis, Monash University 1975
45. Laurent, T. C., Preston, B. N., Sundelöf, L.-O.: Nature *279*, 60 (1979)
46. Preston, B. N. et al.: Nature *287*, 499 (1980)
47. Preston, B. N. et al.: J. Phys. Chem. *87*, 655 (1983)
48. Sundelöf, L.-O.: Anal. Biochem. *127*, 282 (1982); Laurent, T. C. et al.: Anal. Biochem. *127*, 287 (1982)
49. Comper, W. D.: (unpublished)
50. Laurent, T. C. et al.: J. Phys. Chem. *87*, 648 (1983)
51. Preston, B. N. et al.: J. Phys. Chem. *87*, 662 (1983)
52. Comper, W. D. et al.: J. Phys. Chem. *87*, 667 (1983)
53. Comper, W. D., Preston, B. N.: Biochem. Int. *3*, 557 (1981)
54. Nicolis, G., Prigogine, I.: Self-Organization in Non-Equilibrium Systems. From Dissipative Structures to Order Through Fluctuations, New York, John Wiley & Sons Inc. 1977
55. Haken, H.: Synergetics, Berlin—Heidelberg—New York, Springer-Verlag 1977

56. See e.g. Chandrasekhar, S.: Hydrodynamic and Hydromagnetic Stability, London, Oxford Univ. Press 1961
57. Wendt, R. P.: J. Phys. Chem. *66*, 1740 (1962)
58. Reinfelds, G., Gosting, L. J.: J. Phys. Chem. *68*, 2464 (1964)
59. Vitagliano, V. et al.: J. Phys. Chem. *76*, 2050 (1972)
60. See e.g. Sternling, C. V., Scriven, L. E.: A.I.Ch.E. Journal *5*, 514 (1959)
61. Whitmore, R. L.: Brit. J. Appl. Phys. *6*, 239 (1955)
62. Weiland, R. H., McPherson, R. R.: Ind. Eng. Chem. Fundam. **18**, 45 (1979)
63. Comper, W. D., Checkley, G. J., Preston, B. N.: Proc. Aust. Biochem. Soc. *14*, 29 (1981) and unpublished results
64. McDougall, T. J., Turner, J. S.: Nature *299*, 812 (1982); McDougall, T. J.: J. Fluid Mech. *126*, 379 (1983)
65. Turner, J. S.: 2nd Australasian Conf. Heat and Mass Transfer, Univ. of Sydney, p. 1, 1977
66. Anderson, N. G.: Exptl. Cell Res. *9*, 446 (1955)
67. Brakke, M. K.: Arch. Biochem. Biophys. *55*, 175 (1955)
68. Schumaker, V. N. in: Advances in Biological and Medical Physics (eds.) Lawrence, J. H., Gofman, J. W., p. 245, Vol. II, New York, Academic Press 1967
69. Sartory, W. K.: Biopolymers *7*, 251 (1969)
70. Huppert, H. E., Manins, P. C.: Deep. Sea Res. *20*, 315 (1973)
71. Comper, W. D., Preston, B. N.: J. Coll. Int. Sci. (submitted)
72. Ostrach, S. in: Progress in Astronautics and Aeronautics, New York, Amer. Inst. Aeronautics and Astronautics, Vol. 52, p. 3, 1977
73. Karlsson, J. O. in: Molecular Approaches to Neurobiology (ed.) Brown, I. R., p. 131, New York, Academic Press 1982
74. Comper, W. D., Preston, B. N., Austin, L.: Neurochem. Res. (in press)
75. See e.g. Winet, H., Hahn, T. L.: Biorheology *9*, 57 (1972)

H. J. Cantow (Editor)
Received April 21, 1983

Author Index Volumes 1–55

Subject Index

Advances in Polymer Science

Fortschritte der
Hochpolymeren-Forschung

Editors: H.-J. Cantow,
G. Dall'Asta, K. Dušek,
J. D. Ferry, H. Fujita,
M. Gordon, J. P. Kennedy,
W. Kern, S. Okamura,
C. G. Overberger, T. Saegusa,
G. V. Schulz, W. P. Slichter,
J. K. Stille

Springer-Verlag
Berlin
Heidelberg
New York
Tokyo

Volume 43

Polymerizations and Polymer Properties

1982. 94 figures. V, 252 pages
ISBN 3-540-11048-8

Contents: *J. P. Kennedy, V. S. C. Chang, A. Guyot:* Carbocationic Synthesis and Characterization of Polyolefins with Si-H and Si-CI Head Groups. – *A. Fradet, E. Maréchal:* Kinetics and Mechanisms of Polyesterifications. I. Reactions of Diols with Diacids. – *E. Heidemann, W. Roth:* Synthesis and Investigation of Collagen Model Peptides. – *G. K. Elyashevich:* Thermodynamics and Kinetics of Oriental Crystallization of Flexible-Chain Polymers.

Volume 50

Unusual Properties of New Polymers

1983. 32 figures. V, 149 pages
ISBN 3-540-12048-3

Contents: *J. Pitha:* Physiological Activities of Synthetic Analogs of Polynucleotides. – *G. Smets:* Photochromic Phenomena in the Solid Phase. – *D. Wöhrle:* Polymer Square Planar Metal Chelates for Science and Industry/Synthesis, Properties and Applications.

Volume 51

Industrial Developments

1983. 60 figures, 52 tables. Approx. 240 pages
ISBN 3-540-12189-7

Contents: *G. Henrici-Olivé and S. Olivé:* The Chemistry of Carbon Fiber Formation from Polyacrylonitrile. – *V. A. Zakharov, G. D. Bukatov, Y. I. Yermakov:* On the Mechanism of Olefin Polymerization by Ziegler-Natta Catalysts. – *U. Zucchini, G. Cecchin:* Control of Molecular-Weight Distribution in Polyolefins Synthesized with Ziegler-Natta Catalytic Systems. – *F. A. Shutov:* Foamed Polymers. Cellular Structure and Properties.

G. W. Gokel,
S. H. Korzeniowski

Macrocyclic Polyether Syntheses

1982. 89 tables. XVIII, 410 pages
(Reactivity and Structure, Volume 13)
ISBN 3-540-11317-7

Contents: Introduction and General Principles. – The Template Effect. – Syntheses of Oxygen Macrocycles. – Syntheses of Azacrowns. – Crown Esters and Macrocyclic Polyether Lactones. – Miscellaneous Macrocycles. – Open-chained Equivalents of Crown Ethers. – Cryptands and Related Polycyclic Systems. – Author Index. – Subject Index.

This book is the most comprehensive review of macrocyclic polyethers available today. The authors have collected structural information on several thousand existing polyether compounds and their relatives, and have summarized the synthetic approaches which have been used in obtaining them. The availability of such a reference will help facilitate entry of new practitioners into the field and assist those established workers who have long felt need for literature references and for a catalog of extensive tables presenting the structures. It also contains commentary on conflicting data, the specific aims of programs which required the synthesis of certain classes of compounds, and a comparison of methodology where appropriate. Detailed instructions for using the tables complete the work, enhancing its value for researchers in both academia and industry.

Springer-Verlag
Berlin
Heidelberg
New York
Tokyo